Introducing .NET for Apache Spark

Distributed Processing for Massive Datasets

Ed Elliott

Apress®

Introducing .NET for Apache Spark: Distributed Processing for Massive Datasets

Ed Elliott
Sussex, UK

ISBN-13 (pbk): 978-1-4842-6991-6   ISBN-13 (electronic): 978-1-4842-6992-3
https://doi.org/10.1007/978-1-4842-6992-3

Managing Director, Apress Media LLC: Welmoed Spahr
Acquisitions Editor: Jonathan Gennick
Development Editor: Laura Berendson
Coordinating Editor: Jill Balzano

Cover image designed by Freepik (www.freepik.com)

Distributed to the book trade worldwide by Springer Science+Business Media LLC, 1 New York Plaza, Suite 4600, New York, NY 10004. Phone 1-800-SPRINGER, fax (201) 348-4505, e-mail orders-ny@springer-sbm.com, or visit www.springeronline.com. Apress Media, LLC is a California LLC and the sole member (owner) is Springer Science + Business Media Finance Inc (SSBM Finance Inc). SSBM Finance Inc is a **Delaware** corporation.

For information on translations, please e-mail booktranslations@springernature.com; for reprint, paperback, or audio rights, please e-mail bookpermissions@springernature.com.

Apress titles may be purchased in bulk for academic, corporate, or promotional use. eBook versions and licenses are also available for most titles. For more information, reference our Print and eBook Bulk Sales web page at http://www.apress.com/bulk-sales.

Any source code or other supplementary material referenced by the author in this book is available to readers on GitHub via the book's product page, located at www.apress.com/9781484269916. For more detailed information, please visit http://www.apress.com/source-code.

Printed on acid-free paper

To Sarah, Sammy, and Lucy.

Table of Contents

About the Author

Ed Elliott is a data engineer who has been working in IT for 20 years and has focused on data for the last 15 years. He uses Apache Spark at work and has been contributing to the Microsoft .NET for Apache Spark open source project since it was released in 2019. Ed has been blogging and writing since 2014 at his own blog as well as for SQL Server Central and Redgate. He has spoken at a number of events such as SQLBits, SQL Saturday, and the GroupBy conference.

About the Technical Reviewer

Gerald Versluis is a software engineer at Microsoft from the Netherlands. With years of experience working with Azure, ASP.NET, Xamarin, and lots of other .NET technologies, he has been involved in numerous projects big or small.

Not only does he like to code, but he is also passionate about spreading his knowledge – as well as gaining some in the bargain. Gerald involves himself in speaking, providing training sessions, writing blogs (`https://blog.verslu.is`) or articles, recording videos on YouTube (`https://youtube.com/GeraldVersluis`), and contributing to open source projects in his spare time. Twitter: @jfversluis | All handles: `https://jfversluis.dev`

Introduction

From the first time I ever used Apache Spark, I was captivated by how powerful it was and what it could enable. Having come from a traditional Microsoft developer background, C, C++, C#, F#, Microsoft SQL Server, SSIS, and then onto Azure, Azure Data Factory, and SQL Data Warehouse, to be able to use Apache Spark to process large files quickly was fascinating. The less fascinating part was having to decide whether to learn Scala or Python to work with Apache Spark. I studied both languages but missed developing in C# and F#, so when at the launch of the .NET for Apache Spark project, I was particularly excited. I have been closely following the project and have submitted several successful pull requests, which are now included in the bindings.

This book intends to introduce both Apache Spark and the .NET for Apache Spark bindings. If you are new to Apache Spark but have C# or F# experience, or whether you already know Apache Spark but will use the .NET for Apache Spark bindings, then you will be able to use this book to get you started.

In this book, we will start with a step-by-step guide on getting an instance of Apache Spark running on your developer machine, whether on Windows, Linux, or macOS, including a discussion on the crucial points such as the Java version. We will cover in detail how the .NET bindings work. We will then get you up and running with your first .NET for Apache Spark application before explaining the different APIs and taking a look at some example programs.

We will walk through an example batch mode, streaming, and machine learning application that you can follow along. We will also look at how to troubleshoot Apache Spark and, finally, look at how you can make your changes to the .NET for Apache Spark project, including submitting any pull requests back to the core product, if you would like to. Finally, we will look at the support for the delta format, which brings ACID properties to files in a data lake. Exciting times!

PART I

Getting Started

CHAPTER 1

Understanding Apache Spark

Apache Spark is a data analytics platform that has made big data accessible and brings large-scale data processing into the reach of every developer. With Apache Spark, it is as easy to read from a single CSV file on your local machine as it is to read from a million CSV files in a data lake.

An Example

Let us look at an example. The code in Listings 1-1 (C#) and 1-2 (the F# version) reads from a set of CSV files and counts how many records match a specific condition. The code reads all CSV files in a specific path, so the number of files we read from is practically limitless.

Although the examples in this chapter are fully functioning samples, they require a working Apache Spark instance, either locally or on a cluster. We cover setting up Apache Spark in Chapter 2 and running .NET for Apache Spark in Chapter 3.

Listing 1-1. Counting how many rows match a filter in one or a million CSV files in C#

```
using System;
using System.Linq;
using Microsoft.Spark.Sql;
using static Microsoft.Spark.Sql.Functions;
```

© Ed Elliott 2021
E. Elliott, *Introducing .NET for Apache Spark*, https://doi.org/10.1007/978-1-4842-6992-3_1

```
namespace Introduction_CSharp
{
    class Program
    {
        static void Main(string[] args)
        {
            var path = args.FirstOrDefault();

            var spark = SparkSession
                .Builder()
                .GetOrCreate();

            var dataFrame = spark.Read().Option("header", "true").
            Csv(path);
            var count = dataFrame.Filter(Col("name") == "Ed Elliott").
            Count();
            Console.WriteLine($"There are {count} row(s)");
        }
    }
}
```

Listing 1-2. Counting how many rows match a filter in one or a million
CSV files in F#

```
open Microsoft.Spark.Sql

[<EntryPoint>]
let main argv =

    let path = argv.[0]
    let spark = SparkSession.Builder().GetOrCreate()

    spark.Read().Option("header", "true").Csv(path)
     |> fun dataframe -> dataframe.Filter(Functions.Col("name").EqualTo("Ed
        Elliott")).Count()
     |> printfn "There are %d row(s)"

    0
```

Executing either of these programs displays the number of rows matching the filter:

```
» dotnet run --project ./Listing0-1 "/Users/ed/sample-data/1.csv"
There are 1 row(s)
» dotnet run --project ./Listing0-2 "/Users/ed/sample-data/1.csv"
There are 1 row(s)
```

If we use this for a single file, then that is fine, and the code looks quite efficient, but when the same code can run, as is, across a cluster of many nodes and petabytes of data, efficiently, then you can see how powerful Apache Spark can be.

The Core Use Cases

Apache Spark is unique in the world of big data processing in that it allows for data processing, analytics, as well as machine learning. Typically, you can use Apache Spark:

- To transform your data as part of your ETL or ELT data pipelines

- To analyze datasets from one small file to petabytes of data across millions of files

- To create machine learning (ML) applications to enable AI

Transform Your Data

Apache Spark can read and write from any file format or database that is supported by the Java Virtual Machine, so that means we can read from a JDBC connection and write to a file. Apache Spark comes out of the box with the ability to read from a wide range of file formats, such as CSV or Parquet. However, you can always reference additional JAR files to add support for additional file types, for example, the crealytics "spark-excel" plugin (`https://github.com/crealytics/spark-excel`) allows you to read and write from XLSX files in Apache Spark.

To show an example of how powerful Apache Spark is when processing and to show how it really was built for performance from the ground up, I worked on one project where we would read a huge parquet file that contained all the Adobe Clickstream data for a popular, international, website. In our case, the data was one single file that contains all the user's actions on a website; for a well-visited website, the file can be multiple GB and contains a whole range of events, including invalid data. My team was tasked with efficiently reading the entire file of millions of rows and retrieving a minimal

5

subset of one specific action. Before Apache Spark, we would have likely brought the entire file into a database and then filtered out the rows we didn't want or use a tool such as Microsoft's SSIS, which would have read in the entire file. When we implemented this in Apache Spark, we wrote a filter for the specific row type we wanted. Apache Spark read from the file and used predicate pushdown to pass the filter to the driver that read the parquet file, so, at the very earliest opportunity, invalid rows were filtered out. This project demonstrated to us that Apache Spark showed a level of performance and ease of use that our team had not witnessed before.

The code in Listings 1-3 (C#) and 1-4 (F#) will demonstrate how to read from a data source, filter the data to just the rows you require, and show how to write the data out to a new file, which couldn't be more straightforward with Apache Spark.

Listing 1-3. Reading, filtering, and writing data back out again in C#

```csharp
using System;
using Microsoft.Spark.Sql;

namespace TransformingData_CSharp
{
    class Program
    {
        static void Main(string[] args)
        {
            var spark = SparkSession
                .Builder()
                .GetOrCreate();

            var filtered = spark.Read().Parquet("1.parquet")
                .Filter(Functions.Col("event_type") == Functions.Lit(999));

            filtered.Write().Mode("overwrite").Parquet("output.parquet");
            Console.WriteLine($"Wrote: {filtered.Count()} rows");
        }
    }
}

» dotnet run --project ./ Listing0-3
Wrote: 10 rows
```

Listing 1-4. Reading, filtering, and writing data back out again in F#

```fsharp
open Microsoft.Spark.Sql
open System

[<EntryPoint>]
let main argv =

    let writeResults (x:DataFrame) =
        x.Write().Mode("overwrite").Parquet("output.parquet")
        printfn "Wrote: %u rows" (x.Count())

    let spark = SparkSession.Builder().GetOrCreate()
    spark.Read().Parquet("1.parquet")
    |> fun p -> p.Filter(Functions.Col("Event_Type").EqualTo(Functions.
       Lit(999)))
    |> fun filtered -> writeResults filtered

    0 // return an integer exit code

» dotnet run --project ./ Listing0-4
Wrote: 10 rows
```

Analyze Your Data

Apache Spark includes the data analytical abilities you would expect from a database such as aggregation, windowing, and SQL functions, which you can access using the public API such as `data.GroupBy(Col("Name")).Count()`. Interestingly, you can also write Spark SQL, which means you can use SQL queries to access your data. Spark SQL makes Apache Spark available to a much wider audience, which includes developers as well as analysts and data scientists. The ability to access the power of Apache Spark without needing to learn one of Scala, Python, Java, R, and now C# or F# is a compelling feature.

Listings 1-5 and 1-6 show another example where we generate three datasets, union the datasets together, and then aggregate and display the results in .NET, and then in Listing 1-7, we demonstrate the same result, but instead of using .NET code, we pass a SQL query to Apache Spark and execute that query to create a result set we can use; note that there are some Apache Spark environments like Databricks notebooks where we can write just SQL without any application code.

Listing 1-5. Create three datasets, union, aggregate, and count in C#

```csharp
using System;
using Microsoft.Spark.Sql;
using static Microsoft.Spark.Sql.Functions;

namespace TransformingData_CSharp
{
    class Program
    {
        static void Main(string[] args)
        {
            var spark = SparkSession
                .Builder()
                .GetOrCreate();

            var data = spark.Range(100).WithColumn("Name", Lit("Ed"))
                .Union(spark.Range(100).WithColumn("Name", Lit("Bert")))
                .Union(spark.Range(100).WithColumn("Name",
                Lit("Lillian")));

            var counts = data.GroupBy(Col("Name")).Count();
            counts.Show();
        }
    }
}
```

Listing 1-6. Create three datasets, union, aggregate, and count in F#

```fsharp
open Microsoft.Spark.Sql
open System

[<EntryPoint>]
let main argv =
    let spark = SparkSession.Builder().GetOrCreate()
    spark.Range(100L).WithColumn("Name", Functions.Lit("Ed"))
    |> fun d -> d.Union(spark.Range(100L).WithColumn("Name", Functions.
       Lit("Bert")))
```

```
|> fun d -> d.Union(spark.Range(100L).WithColumn("Name", Functions.
   Lit("Lillian")))
|> fun d -> d.GroupBy(Functions.Col("Name")).Count()
|> fun d -> d.Show()

0
```

Finally, in Listing 1-7, we will use Spark SQL to achieve the same result.

Listing 1-7. Create three datasets, union, aggregate, and count in Spark SQL

```
using System;
using Microsoft.Spark.Sql;

namespace TransformingData_SQL
{
    class Program
    {
        static void Main(string[] args)
        {
            var spark = SparkSession
                .Builder()
                .GetOrCreate();

            var data = spark.Sql(@"
                WITH users
                AS (
                    SELECT ID, 'Ed' as Name FROM Range(100)
                    UNION ALL
                    SELECT ID, 'Bert' as Name FROM Range(100)
                    UNION ALL
                    SELECT ID, 'Lillian' as Name FROM Range(100)
                ) SELECT Name, COUNT(*) FROM users GROUP BY Name
            ");
            data.Show();
        }
    }
}
```

The code that is executed by Apache Spark is the same in all three instances and results in the following output:

```
» dotnet run --project ./Listing0-7
+-------+--------+
|   Name|count(1)|
+-------+--------+
|   Bert|     100|
|Lillian|     100|
|     Ed|     100|
+-------+--------+
```

Machine Learning

The last core use case for Apache Spark is to write machine learning (ML) applications. Today, there are quite a few different environments to write ML applications such as Scikit-Learn, TensorFlow, and PyTorch. However, the advantage of using Apache Spark for your ML application is that if you already process your data with Apache Spark, then you get the same familiar API, and more importantly, you can reuse your existing infrastructure.

To see what sort of things you can do in Apache Spark with the ML API, see https://spark.apache.org/docs/latest/ml-guide.html.

.NET for Apache Spark

Apache Spark is written in Scala and runs on the Java Virtual Machine (JVM), but there are a large number of developers whose primary language is C# and, to a lesser extent, F#. The .NET for Apache Spark project aims to bring the full capabilities of Apache Spark to .NET developers. Microsoft started the project as an open source project, developing in the open and accepting pull requests, issues, and feature requests.

The .NET for Apache Spark project provides an interop layer between the .NET CLI code and the JVM. The way this works is that there is a Java class, written in Scala; the Java class called the DotnetRunner creates a TCP socket, and then the DotnetRunner runs a dotnet program, **your program** which creates a SparkSession. The SparkSession makes a connection to the TCP socket and forwards requests to the JVM and returns the response. You can think of the .NET for Apache Spark library as a proxy between your .NET code and the JVM.

The Microsoft team made an important early decision, which affects how we can use Apache Spark from .NET. Apache Spark originally started with what is called the RDD API. The RDD API allows users to access the underlying data structure used by Apache Spark. When Apache Spark version 2.0 was released, it included a new DataFrame API. The DataFrame API had several additional benefits such as a new `"catalyst"` optimizer, which meant that it was much more efficient to use the DataFrame API than the original RDD API. Letting Apache Spark optimize the query, rather than trying to optimize the calls yourself using the RDD API, was also a lot simpler. The DataFrame API brought performance parity to Python and R, and now .NET. The previous RDD API was considerably faster for Scala or Java code than it was with Python. With the new DataFrame API, it was just as fast, in most cases, for Python or R code as it was with Scala and Java code.

The Microsoft team decided only to provide support for the new DataFrame API, which means it isn't possible, today, to use the RDD API from .NET for Apache Spark. I honestly do not see this as a significant issue, and it certainly is not a blocker for the adoption of .NET for Apache Spark. This condition of only supporting the later API flows through to the ML library, where there are two APIs for ML, MLLib and ML. The Apache Spark team deprecated MLLib in favor of the ML library, so in .NET for Apache Spark, we are also only implementing the ML version of the API.

Feature Parity

The .NET for Apache Spark project was first released to the public in April 2019 and included a lot of the core functionality that is also available in Apache Spark. However, there was quite a lot of functionality missing, even from the DataFrame API, and that is ignoring the APIs which are likely not going to be implemented, such as the RDD API. In the time since the initial release, the Microsoft team and outside contributors have increased the amount of functionality. In the meantime, the Apache Spark team has also released more functionality, so in some ways, the Microsoft project is playing catch-up with the Apache team, so not all functionality is currently available in the .NET project. Over the last year and a bit, the gap has been closing, and I fully expect over the next year or so the gap to get smaller and smaller, and feature parity will exist at some point.

If you are trying to use the .NET for Apache Spark project and some functionality is missing that is a blocker for you, there are a couple of options that you can take to implement the missing functionality, and I cover this in Appendix B.

Summary

Apache Spark is a compelling data processing project that makes it almost too simple to query large distributed datasets. .NET for Apache Spark brings that power to .NET developers, and I, for one, am excited by the possibility of creating ETL, ELT, ML, and all sorts of data processing applications using C# and F#.

CHAPTER 2

Setting Up Spark

So that we can develop a .NET for Apache Spark application, we need to install Apache Spark on our development machines and then configure .NET for Apache Spark so that our application executes correctly. When we run our Apache Spark application in production, we will use a cluster, either something like a YARN cluster or using a fully managed environment such as Databricks. When we develop applications, we use the same version of Apache Spark locally as we would when we run against a cluster of many machines. Having the same version on our development machines means that when we develop and test the code, we can be confident that the code that runs in production is the same.

In this chapter, we will go through the various components that we need to have running correctly; Apache Spark is a Java application so we will need to install and configure the correct version of Java and then download and configure Apache Spark. Only when we have the correct version of Java and Apache Spark running are we able to write a .NET application, either in C# or F# that executes on Apache Spark.

Choosing Your Software Versions

In this section, we are going to start by helping you choose which version of Apache Spark and which version of Java you should use. Even though it seems like it should be a straightforward choice, there are some specific requirements, and getting this correct is critical to getting off to a smooth start.

Choosing a Version of Apache Spark

In this section, we will look at how to choose a version of Apache Spark. Apache Spark is an actively developed open source project, and new releases happen often, sometimes even multiple times a month. However, the .NET for Apache Spark project does not support every version, either because it will not support it or because the development team hasn't yet added.

13

© Ed Elliott 2021
E. Elliott, *Introducing .NET for Apache Spark*, https://doi.org/10.1007/978-1-4842-6992-3_2

When we run a .NET for Apache Spark application, we need to understand that we need the .NET code, which runs on a specific version of the .NET Framework or .NET Core. The .NET for Apache Spark code is compatible with a limited set of versions of Apache Spark, and depending on which version of Apache Spark you have, you will either need Java 8 or Java 11.

To help choose the version of the components that you need, go to the home page of the .NET for Apache Spark project, `https://github.com/dotnet/spark`, and there is a section "Supported Apache Spark"; the current .NET for Apache Spark version `"v1.0.0"` supports these versions of Apache Spark:

- 2.3.*

- 2.4.0

- 2.4.1

- 2.4.3

- 2.4.4

- 2.4.5

- 3.0.0

Note that 2.4.2 is not supported, and 3.0.0 of Apache Spark was supported when .NET for Apache Spark v1.0.0 was released in October 2020. Where possible, you should aim for the highest version of both projects that you can, and, today, in November 2020, I would start a new project with .NET for Apache Spark v1.0.0 and Apache Spark version 3.0. Unfortunately, any concrete advice we write here will quickly get out of date. Between the time of writing this and reviewing the chapter, the advice changed from using .NET for Apache Spark version v0.12.1 and v1.0.0.

Once you have selected a version of the Apache Spark code to use, visit the release notes for that version, such as `https://spark.apache.org/docs/3.0.0/` or `https://spark.apache.org/docs/3.0.0/`. The release notes include details of which version of the Java VM is supported. If you try and run on a version of the JVM that is not supported, then your application will fail, so you do need to take care here.

When you download Apache Spark, you have a few options. You can download the source code and compile it by yourself, which we do not cover here, but you can get instructions on how to build from source from *https://spark.apache.org/docs/latest/building-spark.html*. You can also choose to either download with a pre-built

Hadoop or without Hadoop, but you would then need to provide your own Hadoop installation. Typically, for a development machine, I would download the version of Apache Spark with pre-built Hadoop, so you do not have to maintain an instance of Hadoop. There are cases though where you would want to use the version without Hadoop and install Hadoop separately. One example is if you, from your development instance, want to read and write files to something like Azure Data Lake storage, then you will need a separate implementation of Hadoop as the pre-built one. At the time of writing, the pre-built Hadoop implementation did not support the Azure Data Lake protocol. In practice, you will likely find that you develop against local files and only require reading and writing to things like Azure Data Lake storage when you are running on a cluster.

To summarize:

- **Choose a .NET for Apache Spark version** – Ideally, get the latest.

- **Choose an Apache Spark Version** – The latest supported by .NET for Apache Spark.

- **Choose a Java VM version** – Choose the latest supported by your choice of Apache Spark.

- **Choose a download with or without Hadoop** – Unless you have a specific requirement to use a separate Hadoop, use the version of Apache Spark with Hadoop pre-built.

In the next section, we will explore how to choose a version of Java, which is not always as straightforward as it first appears.

Choosing a Java Version

Sun Microsystems initially developed Java, which was available to anyone for free. In April 2019, Oracle changed the way that Java was licensed, and so from April 2019, the version of Java from Oracle over version 8 is not free for production or nonpersonal use anymore. To add to the confusion, Oracle also releases a version of Java called the OpenJDK, which does not have these restrictions; many people are choosing to use the OpenJDK version of Java over the Oracle version of Java. To read more about these licensing changes, see `www.oracle.com/java/technologies/javase/jdk-faqs.html`.

When talking about Java versions, two different schemes point to the same logical version; Java 1.8 and Java 8, although look very different, are the same version.

What you do get with the Oracle JDK over the Oracle OpenJDK is support, which is an essential requirement in many organizations. The top two answers on this Stack Overflow post describe the issue and help guide your choice on which flavor of Java to use: https://stackoverflow.com/questions/52431764/difference-between-openjdk-and-adoptium-adoptopenjdk.

Configuring Apache Spark and .NET for Apache Spark on macOS

In this section, we will cover how to get a local instance of Apache Spark running development machine; once we have a working installation with a version that is supported by .NET, we can create our first .NET application in the next chapter. In later sections, we will look at how to configure Apache Spark on Windows and then Linux.

Configuring Already Installed Java

Before you install Java, it is worth checking that you do not already have the correct version of Java installed and configured or that you have the correct version installed but a separate version configured. To see which versions of Java, if any, you have on your macOS, run /usr/libexec/java_home -V. In this case, the output shows the Java 8 and 13 JDKs:

```
» /usr/libexec/java_home -V
Matching Java Virtual Machines (2):
    13.0.1, x86_64:    "OpenJDK 13.0.1"    /Library/Java/
    JavaVirtualMachines/openjdk-13.0.1.jdk/Contents/Home
    1.8.0_232, x86_64:    "AdoptOpenJDK 8"    /Library/Java/
    JavaVirtualMachines/adoptopenjdk-8.jdk/Contents/Home
```

In my case, I need Java 8 so I can run Apache Spark "3.0.0," and so I check which version of the two Java instances is set to default:

```
» java -version
openjdk version "1.8.0_232"
OpenJDK Runtime Environment (AdoptOpenJDK)(build 1.8.0_232-b09)
OpenJDK 64-Bit Server VM (AdoptOpenJDK)(build 25.232-b09, mixed mode)
```

If the incorrect version were chosen, then I would need to update my JAVA_HOME environment variable using the tool "/usr/libexec/java_home" and passing that tool the version of Java that I want to use:

```
» export JAVA_HOME=$(/usr/libexec/java_home -v 13)
» java -version
openjdk version "13.0.1" 2019-10-15
OpenJDK Runtime Environment (build 13.0.1+9)
OpenJDK 64-Bit Server VM (build 13.0.1+9, mixed mode, sharing)
```

or

```
» export JAVA_HOME=$(/usr/libexec/java_home -v 1.8)
» java -version
openjdk version "1.8.0_232"
OpenJDK Runtime Environment (AdoptOpenJDK)(build 1.8.0_232-b09)
OpenJDK 64-Bit Server VM (AdoptOpenJDK)(build 25.232-b09, mixed mode)
```

Installing Java

If you do not have a version of Java that Apache Spark can use, we will need to download a version of the JDK. In this section, we will download and install the AdoptOpenJDK 8 JDK, so we go to the releases page (https://adoptopenjdk.net/releases.html) and choose

- OpenJDK 8 (LTS)
- HotSpot
- macOS
- JDK .pkg

These options download a pkg file that you can then install using the GUI or the command line `installer -pkg jdk.pkg`. If you download the package, the package will install and configure Java. If you start a new terminal and run `java -version`, check that the correct version now shows; if it does not show the correct version, then follow the steps in the previous section.

Downloading and Configuring Apache Spark

Now that you have the correct working version of Java, you can download Apache Spark. Go to the home page (`https://spark.apache.org/downloads.html`), choose the version you want and the package type, and then download; I downloaded and copied mine to my home directory and then ran

```
» tar -xvf spark-3.0.0-bin-hadoop2.7.tgz
```

This command extracted the files to ~/ spark-3.0.0-bin-hadoop2.7/, so the next step is to set up a couple of environment variables:

- SPARK_HOME

- PATH

We set `SPARK_HOME` to the directory that we just extracted, and I would update my .zshrc with the new folder and also update my PATH variable in .zshrc to include "$SPARK_HOME/bin"; make sure you set $SPARK_HOME before you update your path.

Test

To verify that we have a working installation of Apache Spark, run `spark-shell`, which is a REPL for running commands. If you run `spark-shell`, you should see the spark logo displayed and a command prompt where you can run spark commands.

```
» spark-shell
Setting default log level to "WARN".
To adjust logging level use sc.setLogLevel(newLevel). For SparkR, use
setLogLevel(newLevel).
Spark context Web UI available at http://localhost:4040
Spark context available as 'sc' (master = local[*], app id =
local-1595401509136).
```

```
Spark session available as 'spark'.
Welcome to
      ____              __
     / __/__  ___ _____/ /__
    _\ \/ _ \/ _ `/ __/  '_/
   /___/ .__/\_,_/_/ /_/\_\   version 3.0.0
      /_/

Using Scala version 2.12.10 (OpenJDK 64-Bit Server VM, Java 1.8.0_262)
Type in expressions to have them evaluated.
Type :help for more information.

scala>
```

If you get the scala> prompt, then that is a great sign that everything is working, but let us see if we can run a spark command to do some local processing:

```
scala> spark.sql("select * from range(10)").withColumn("ID2", col("ID")).show
+---+---+
| id|ID2|
+---+---+
|  0|  0|
|  1|  1|
|  2|  2|
|  3|  3|
|  4|  4|
|  5|  5|
|  6|  6|
|  7|  7|
|  8|  8|
|  9|  9|
+---+---+
```

What we did here was to create a DataFrame by running some spark SQL to call the range function. The range function creates as many rows as we ask it, then we used withColumn to create a second column with the same value that was in the first column. Finally, we used show to display the contents of the DataFrame.

Exiting the spark-shell REPL is like exiting VIM; use :q.

Overriding Default Config

To configure Apache Spark, we can use the configuration files that are in $SPARK_
HOME/conf; these are a set of text files that control how Apache Spark works. When you
first download Apache Spark, there are only template config files, no actual config files:

```
~/spark-3.0.0-bin-hadoop2.7/conf » ls
docker.properties.template    metrics.properties.template  spark-env.
sh.template
fairscheduler.xml.template    slaves.template
log4j.properties.template     spark-defaults.conf.template
```

If we want to configure any of the config files, we should first copy the template file to
the actual file by removing ".template" from the end of the filenames:

```
» cp ./spark-defaults.conf.template ./spark-defaults.conf
» cp ./log4j.properties.template ./log4j.properties
```

You can then use your favorite editor to open the config files and edit them. For a
development instance of Apache Spark, I would change the log4j section for console
logging to just showing errors; showing warnings creates a lot of output that we can
normally ignore. If you change the default file contents which looks like:

```
# Set everything to be logged to the console
log4j.rootCategory=INFO, console
log4j.appender.console=org.apache.log4j.ConsoleAppender
log4j.appender.console.target=System.err
log4j.appender.console.layout=org.apache.log4j.PatternLayout
log4j.appender.console.layout.ConversionPattern=%d{yy/MM/dd HH:mm:ss} %p
%c{1}: %m%n
```

to

```
# Set only errors to be logged to the console
log4j.rootCategory=ERROR, console
log4j.appender.console=org.apache.log4j.ConsoleAppender
log4j.appender.console.target=System.err
log4j.appender.console.layout=org.apache.log4j.PatternLayout
log4j.appender.console.layout.ConversionPattern=%d{yy/MM/dd HH:mm:ss}
%p %c{1}: %m%n
```

that is usually a good start. In the spark-defaults.conf file, it is often a good idea to configure how much memory Apache Spark can use on your development machine; my machine, for instance, has 16GB of RAM, so I configure Apache Spark to use 6GB for the executor memory. I also use the Delta Lake library from Databricks quite extensively, so by adding it to my default config, I get access to it every time and do not need to remember to start Apache Spark with the additional library:

```
spark.executor.memory=6g
spark.jars.packages=io.delta:delta-core_2.12:0.7.0
```

Once I have changed the config files, it is always worth rerunning spark-shell to verify that the changes have worked. This time, when I run spark-shell, I can see that the logging is less verbose and also that my delta package has been loaded:

```
» ./spark-shell
Ivy Default Cache set to: /Users/ed/.ivy2/cache
The jars for the packages stored in: /Users/ed/.ivy2/jars
:: loading settings :: url = jar:file:/Users/ed/spark-3.0.0-bin-without-
hadoop/jars/ivy-2.4.0.jar!/org/apache/ivy/core/settings/ivysettings.xml
io.delta#delta-core_2.12 added as a dependency
:: resolving dependencies :: org.apache.spark#spark-submit-parent-4987d518-
30f9-4696-ac0e-1b20ed99f224;1.0
        confs: [default]
        found io.delta#delta-core_2.12;0.7.0 in central
        found org.antlr#antlr4;4.7 in central
        found org.antlr#antlr4-runtime;4.7 in local-m2-cache
        found org.antlr#antlr-runtime;3.5.2 in central
        found org.antlr#ST4;4.0.8 in central
        found org.abego.treelayout#org.abego.treelayout.core;1.0.3 in spark-
        list
        found org.glassfish#javax.json;1.0.4 in central
        found com.ibm.icu#icu4j;58.2 in central
:: resolution report :: resolve 210ms :: artifacts dl 8ms
        :: modules in use:
        com.ibm.icu#icu4j;58.2 from central in [default]
```

```
io.delta#delta-core_2.12;0.7.0 from central in [default]
org.abego.treelayout#org.abego.treelayout.core;1.0.3 from spark-list
in [default]
org.antlr#ST4;4.0.8 from central in [default]
org.antlr#antlr-runtime;3.5.2 from central in [default]
org.antlr#antlr4;4.7 from central in [default]
org.antlr#antlr4-runtime;4.7 from local-m2-cache in [default]
org.glassfish#javax.json;1.0.4 from central in [default]
```

		modules				artifacts	
	conf	number	search	dwnlded	evicted	number	dwnlded
	default	8	0	0	0	8	0

```
:: retrieving :: org.apache.spark#spark-submit-parent-4987d518-30f9-4696-
ac0e-1b20ed99f224
        confs: [default]
        0 artifacts copied, 8 already retrieved (0kB/7ms)
```

If you can run the Apache Spark REPL and can run spark commands, then it is very likely that you will be able to run your .NET for Apache Spark application when we create our first application in the next chapter.

Configuring Apache Spark and .NET for Apache Spark on Windows

In this section, we will cover how to get a local instance of Apache Spark running development machine; once we have a working installation with a version that is supported by .NET, we can create our first .NET application in the next chapter.

Configuring Already Installed Java

Before you install Java, it is worth checking that you do not already have the correct version of Java installed and configured or that you have the correct version installed but a separate version configured. To see which versions of Java, if any, you have on your Windows machine, have a look which `java.exe` files exist in your `program files` directory. In this case, the output shows the Java 8 and 11 JDKs:

```
C:\>cd %ProgramFiles%

C:\Program Files>attrib java.exe /s
A                       C:\Program Files\AdoptOpenJDK\jdk-11.0.8.10-hotspot\
                        bin\java.exe
A                       C:\Program Files\AdoptOpenJDK\jdk-8.0.262.10-hotspot\
                        bin\java.exe
A                       C:\Program Files\AdoptOpenJDK\jdk-8.0.262.10-hotspot\
                        jre\bin\java.exe
```

You have a choice about whether you set the correct Java version for the entire system or whether you set the version of Java every time you need to run your application. There isn't a hard and fast rule about this choice, but I typically set the version of Java at runtime, unless I am 100% working on a single version of the Java VM.

To set the version of Java at runtime, you need to set two environment variables: JAVA_HOME and PATH. JAVA_HOME points to the folder that is the root folder for the Java version, and PATH includes the path to the folder where java.exe exists:

```
C:\Program Files>SET JAVA_HOME=C:\Program Files\AdoptOpenJDK\jdk-
8.0.262.10-hotspot

C:\Program Files>SET PATH=%JAVA_HOME%\bin;%PATH%
```

Then you must use the command prompt, which you set the environment variables in to start your application that uses the Java VM.

If you would prefer to set the version for your entire system, which is simpler if you can always use the same version, you should go to Windows Settings and search for "Edit the system environment variables" and use the control panel applet to create a "System Variable" for JAVA_HOME that points to your java directory and add the path to the java bin folder that contains java.exe to the beginning of the PATH environment variable.

When you have configured your environment variables, either in a console session or system-wide using the Windows Settings dialog, which requires a reboot, you can test that you have the correct version of Java configured by running java -version:

```
C:\Program Files>java -version
openjdk version "1.8.0_262"
OpenJDK Runtime Environment (AdoptOpenJDK)(build 1.8.0_262-b10)
OpenJDK 64-Bit Server VM (AdoptOpenJDK)(build 25.262-b10, mixed mode).
```

If you decide to configure JAVA_HOME and PATH every time you run your application, it is useful to create a batch file that does the work for you, so you just call that before running your application.

Installing Java

If you do not have a version of Java that Apache Spark can use, we will need to download a version of the JDK. In this section, we will download and install the AdoptOpenJDK 8 JDK, so we go to the releases page (https://adoptopenjdk.net/releases.html) and choose

- OpenJDK 8 (LTS)
- HotSpot
- Windows
- JDK .msi

These options download an MSI file that you can then install. If you download the package, the package will install and configure Java. If you start a new terminal and run java -version, check that the correct version now shows; if it does not show the correct version, then follow the steps in the previous section.

Downloading and Configuring Apache Spark

Now that you have the correct working version of Java, you can download Apache Spark from the home page (https://spark.apache.org/downloads.html), choose the version you want and the package type, and then download.

When Apache Spark has finished downloading, because the file is a gzipped tar archive, you will not be able to extract it unless you use something like WinZip or 7-Zip. We will use 7-Zip first to decompress the tar archive:

```
>"c:\Program Files\7-Zip\7z.exe" x spark-3.0.0-bin-hadoop2.7.gz
```

7-Zip will then create a file called, in this case, spark-3.0.0-bin-hadoop2.7, which is a tar archive; we can then use 7-Zip again to get to the actual files using

```
>"c:\Program Files\7-Zip\7z.exe" x spark-3.0.0-bin-hadoop2.7 -ospark
```

We end up with a directory called spark with the actual folder we want inside that folder. If you switch to the actual spark directory and run dir, you should see these files and folders:

```
>dir
 Volume in drive C is Windows
 Volume Serial Number is C024-E5C2

 Directory of C:\Users\ed\Downloads\spark\spark-3.0.0-bin-hadoop2.7

05/30/2020  12:02 AM    <DIR>          .
05/30/2020  12:02 AM    <DIR>          ..
05/30/2020  12:02 AM    <DIR>          bin
05/30/2020  12:02 AM    <DIR>          conf
05/30/2020  12:02 AM    <DIR>          data
05/30/2020  12:02 AM    <DIR>          examples
05/30/2020  12:02 AM    <DIR>          jars
05/30/2020  12:02 AM    <DIR>          kubernetes
05/30/2020  12:02 AM            21,371 LICENSE
05/30/2020  12:02 AM    <DIR>          licenses
05/30/2020  12:02 AM            42,919 NOTICE
05/30/2020  12:02 AM    <DIR>          python
05/30/2020  12:02 AM    <DIR>          R
05/30/2020  12:02 AM             3,756 README.md
05/30/2020  12:02 AM               187 RELEASE
05/30/2020  12:02 AM    <DIR>          sbin
05/30/2020  12:02 AM    <DIR>          yarn
```

The next step is to move the spark folder to the location you would like to keep it. You can put the code anywhere you choose; however, I tend to create a c:\spark folder and copy each version inside that folder. In this case, for Apache Spark 3.0.0, it means my spark folder will be c:\spark\spark-3.0.0-bin-hadoop2.7. As with installing and configuring Java, we also need to create an environment variable called SPARK_HOME and update PATH to include the path to the bin folder inside the spark-3.0.0-bin-hadoop2.7 directory. If you will only ever use one version of Apache Spark at a time, then you can change the system environment variables by using Windows Settings and searching for "Edit the system environment variables" and use the control panel applet to create a "System Variable" for SPARK_HOME and edit the environment variable for PATH to include the bin folder we extracted earlier.

If you do use the idea of creating a c:\spark folder, then that is a good place to keep a batch script to configure each version of Apache Spark that you wish to use; for each version of Apache Spark, on Windows, I create a file called c:\spark\version.cmd, so, in this case, c:\spark\3-0-0.cmd, and include the Java and Apache Spark environment variables:

```
SET JAVA_HOME=C:\Program Files\AdoptOpenJDK\jdk-8.0.262.10-hotspot
SET SPARK_HOME=c:\spark\spark-3.0.0-bin-hadoop2.7
SET PATH=%JAVA_HOME%\bin;%SPARK_HOME%\bin;%PATH%
```

Winutils

When you run on Windows, the version of Hadoop that is pre-built, which allows access to read and write files, is missing some core functionality. To read or write files, we need an additional exe call winutils. You can either compile it manually from the Hadoop source or download a pre-built one. If you would like to build from the Hadoop source, then see the instructions here: https://github.com/steveloughran/winutils. If you decide to download a pre-built one, then visit the page (https://github.com/cdarlint/winutils) and choose a version of winutils for your version of Hadoop. In the preceding example, we downloaded spark-3.0.0-bin-hadoop2.7.gz, which is Apache Spark version 3.0.0 and Hadoop version 2.7; the current latest version for Hadoop 2.7 in the winutils repo on GitHub, today, is 2.7.7. If you go into that folder, you can download winutils.exe. You do not need any other file for Apache Spark. The easiest way to configure your Apache Spark instance is to copy your winutils.exe file to your

%SPARK_HOME%\bin directory and set an additional environment variable called HADOOP_ HOME either using the system environment variable if you are only going to use the single version of Apache Spark or by including it in your version-specific batch script:

```
SET JAVA_HOME=C:\Program Files\AdoptOpenJDK\jdk-8.0.262.10-hotspot
SET SPARK_HOME=c:\spark\spark-3.0.0-bin-hadoop2.7
SET HADOOP_HOME=%SPARK_HOME%
SET PATH=%JAVA_HOME%\bin;%SPARK_HOME%\bin;%PATH%
```

Test

To verify that we have a working installation of Apache Spark, run spark-shell, which is a REPL for running commands. If you run spark-shell, you should see the spark logo displayed and a command prompt where you can run spark commands. The first time I ever ran Java and Apache Spark, I also had to allow access to the Windows Firewall.

```
» spark-shell
Using Spark's default log4j profile: org/apache/spark/log4j-defaults.
properties
Setting default log level to "WARN".
To adjust logging level use sc.setLogLevel(newLevel). For SparkR, use
setLogLevel(newLevel).
Spark context Web UI available at http://localhost:4040
Spark context available as 'sc' (master = local[*], app id =
local-1595570221598).
Spark session available as 'spark'.
Welcome to
      ____              __
     / __/__  ___ _____/ /__
    _\ \/ _ \/ _ `/ __/  '_/
   /___/ .__/\_,_/_/ /_/\_\   version 3.0.0
      /_/

Using Scala version 2.12.10 (OpenJDK 64-Bit Server VM, Java 1.8.0_262)
Type in expressions to have them evaluated.
Type :help for more information

scala>
```

If you get the `scala>` prompt, then that is a great sign that everything is working, but let us see if we can run a spark command to do some local processing:

```
scala> spark.sql("select * from range(10)").withColumn("ID2", col("ID")).
show
+---+---+
| id|ID2|
+---+---+
|  0|  0|
|  1|  1|
|  2|  2|
|  3|  3|
|  4|  4|
|  5|  5|
|  6|  6|
|  7|  7|
|  8|  8|
|  9|  9|
+---+---+
```

What we did here was to create a `DataFrame` by running some spark SQL to call the `range` function. The `range` function creates as many rows as we ask it, then we used withColumn to create a second column with the same value that was in the first column. Finally, we used `show` to display the contents of the `DataFrame`.

Exiting the `spark-shell` REPL is like exiting VIM; use `:q`.

When you exit the REPL on Windows, by default, the logging level is configured to show an error that can be ignored. If you see an error such as "ERROR ShutdownHookManager: Exception while deleting Spark temp dir", we will configure the logging to hide this benign error in the next section.

Overriding Default Config

To configure Apache Spark, we can use the configuration files that are in the %SPARK_HOME%\conf folder. These are a set of text files that control how Apache Spark works. When you first download Apache Spark, there are only template config files, no actual config files:

```
C:\spark\spark-3.0.0-bin-hadoop2.7\conf>dir
 Volume in drive C is Windows
 Volume Serial Number is C024-E5C2

 Directory of C:\spark\spark-3.0.0-bin-hadoop2.7\conf

05/30/2020  12:02 AM    <DIR>          .
05/30/2020  12:02 AM    <DIR>          ..
05/30/2020  12:02 AM               996 docker.properties.template
05/30/2020  12:02 AM             1,105 fairscheduler.xml.template
05/30/2020  12:02 AM             2,025 log4j.properties.template
05/30/2020  12:02 AM             7,801 metrics.properties.template
05/30/2020  12:02 AM               865 slaves.template
05/30/2020  12:02 AM             1,292 spark-defaults.conf.template
05/30/2020  12:02 AM             4,221 spark-env.sh.template
```

If we want to configure any of the config files, we should first copy the template file to the actual file, we will need spark-defaults.conf and log4j.properties:

```
C:\spark\spark-3.0.0-bin-hadoop2.7\conf>copy spark-defaults.conf.template
spark-defaults.conf
        1 file(s) copied.
C:\spark\spark-3.0.0-bin-hadoop2.7\conf>copy log4j.properties.template
log4j.properties
        1 file(s) copied.
```

You can then use your favorite editor to open the config files and edit them. For a development instance of Apache Spark, I would change the log4j section for console logging to just showing errors; showing warnings creates a lot of output that we can normally ignore. If you change the default file contents which looks like:

```
# Set everything to be logged to the console
log4j.rootCategory=INFO, console
log4j.appender.console=org.apache.log4j.ConsoleAppender
log4j.appender.console.target=System.err
log4j.appender.console.layout=org.apache.log4j.PatternLayout
log4j.appender.console.layout.ConversionPattern=%d{yy/MM/dd HH:mm:ss} %p
%c{1}: %m%n
```

to

```
# Set only errors to be logged to the console
log4j.rootCategory=ERROR, console
log4j.appender.console=org.apache.log4j.ConsoleAppender
log4j.appender.console.target=System.err
log4j.appender.console.layout=org.apache.log4j.PatternLayout
log4j.appender.console.layout.ConversionPattern=%d{yy/MM/dd HH:mm:ss} %p
%c{1}: %m%n
```

We also need to ensure these two lines are present in the log4j.properties file; otherwise, we see an error every time we exit Apache Spark which does not cause any side effects, just causes confusion and doubt:

```
log4j.logger.org.apache.spark.util.ShutdownHookManager=OFF
log4j.logger.org.apache.spark.SparkEnv=ERROR
```

That is usually a good start. In the spark-defaults.conf file, it is often a good idea to configure how much memory Apache Spark can use on your development machine; my machine, for instance, has 16GB of RAM, so I configure Apache Spark to use 6GB for the executor memory. I also use the Delta Lake library from Databricks quite extensively, so by adding it to my default config, I get access to it every time and do not need to remember to start Apache Spark with the additional library:

```
spark.executor.memory=6g
spark.jars.packages=io.delta:delta-core_2.12:0.7.0
```

Once I have changed the config files, it is always worth rerunning spark-shell to verify that the changes have worked. This time, when I run spark-shell, I can see that the logging is less verbose and also that my delta package has been loaded:

```
C:\spark\spark-3.0.0-bin-hadoop2.7\conf>spark-shell
Ivy Default Cache set to: C:\Users\ed\.ivy2\cache
The jars for the packages stored in: C:\Users\ed\.ivy2\jars
:: loading settings :: url = jar:file:/C:/spark/spark-3.0.0-bin-hadoop2.7/
    jars/ivy-2.4.0.jar!/org/apache/ivy/core/settings/ivysettings.xml
io.delta#delta-core_2.12 added as a dependency
:: resolving dependencies :: org.apache.spark#spark-submit-parent-01c3fcde-
    c454-4d1c-a62a-4a6a55bc3838;1.0
```

```
          confs: [default]
          found io.delta#delta-core_2.12;0.7.0 in central
          found org.antlr#antlr4;4.7 in central
          found org.antlr#antlr4-runtime;4.7 in central
          found org.antlr#antlr-runtime;3.5.2 in central
          found org.antlr#ST4;4.0.8 in central
          found org.abego.treelayout#org.abego.treelayout.core;1.0.3 in
          central
          found org.glassfish#javax.json;1.0.4 in central
          found com.ibm.icu#icu4j;58.2 in central
:: resolution report :: resolve 1390ms :: artifacts dl 32ms
          :: modules in use:
          com.ibm.icu#icu4j;58.2 from central in [default]
          io.delta#delta-core_2.12;0.7.0 from central in [default]
          org.abego.treelayout#org.abego.treelayout.core;1.0.3 from central
          in [default]
          org.antlr#ST4;4.0.8 from central in [default]
          org.antlr#antlr-runtime;3.5.2 from central in [default]
          org.antlr#antlr4;4.7 from central in [default]
          org.antlr#antlr4-runtime;4.7 from central in [default]
          org.glassfish#javax.json;1.0.4 from central in [default]
          ---------------------------------------------------------------------
          |                  |           modules        ||   artifacts  |
          |       conf       | number| search|dwnlded|evicted|| number|dwnlded|
          ---------------------------------------------------------------------
          |     default      |   8   |   0   |   0   |   0   ||   8   |   0   |
          ---------------------------------------------------------------------
:: retrieving :: org.apache.spark#spark-submit-parent-01c3fcde-c454-4d1c-
a62a-4a6a55bc3838
          confs: [default]
          0 artifacts copied, 8 already retrieved (0kB/31ms)
20/07/24 06:10:26 WARN NativeCodeLoader: Unable to load native-hadoop
library for your platform... using builtin-java classes where applicable
Setting default log level to "WARN".
```

To adjust logging level use sc.setLogLevel(newLevel). For SparkR, use
setLogLevel(newLevel).
Spark context Web UI available at http:// win10.jfdetya2p5vexax3rmqk4rrt4a.
ax.internal.cloudapp.net:4040
Spark context available as 'sc' (master = local[*], app id =
local-1595571047482).
Spark session available as 'spark'.
Welcome to

```
      ____              __
     / __/__  ___ _____/ /__
    _\ \/ _ \/ _ `/ __/  '_/
   /___/ .__/\_,_/_/ /_/\_\   version 3.0.0
      /_/
```

Using Scala version 2.12.10 (OpenJDK 64-Bit Server VM, Java 1.8.0_262)
Type in expressions to have them evaluated.
Type :help for more information.

scala>

If you can run the Apache Spark REPL and can run spark commands, then it is very
likely that you will be able to run your .NET for Apache Spark application when we create
our first application in the next chapter.

Configuring Apache Spark and .NET for Apache Spark on Linux (Ubuntu)

In this section, we will cover how to get a local instance of Apache Spark running
development machine; once we have a working installation with a version that is
supported by .NET, we can create our first .NET application in the next chapter.

Configuring Already Installed Java

Before you install Java, it is worth checking that you do not already have the correct
version of Java installed and configured or that you have the correct version installed but
a separate version configured. To see which versions of Java, if any, you have on Ubuntu,
use the update-alternatives tool:

```
$ sudo update-alternatives --config java
There are 2 choices for the alternative java (providing /usr/bin/java).
  Selection      Path
Priority    Status
------------------------------------------------------------
* 0              /usr/lib/jvm/java-11-openjdk-amd64/bin/java
1111         auto mode
  1              /usr/lib/jvm/java-11-openjdk-amd64/bin/java
1111         manual mode
  2              /usr/lib/jvm/java-8-openjdk-amd64/jre/bin/java
1081         manual mode
```

update-alternatives shows we have Java 11 and Java 8 installed. However, Java 11 is the default, which does not support the current latest version of .NET for Apache Spark, which requires Java 8.

You can either choose to change the default by passing in the selection number for the Java 8 runtime, or you can create a script you call before you run your Apache Spark application that always configures the correct version of Java:

```
#!/bin/bash

export JAVA_HOME=$(update-java-alternatives --list java-1.8.0-openjdk-amd64
| awk '{print $3}')
export PATH=$JAVA_HOME/bin:$PATH
```

After we create this shell file, we can source it and then run java -version to check whether the script has worked successfully:

```
$ source ./spark.sh
$ java -version
openjdk version "1.8.0_252"
OpenJDK Runtime Environment (build 1.8.0_252-8u252-b09-1~18.04-b09)
OpenJDK 64-Bit Server VM (build 25.252-b09, mixed mode)
```

In this case, we now see that the Java we have set in the terminal session is Java 8, which is correct.

Installing Java

If you do not have a version of Java that Apache Spark can use, we will need to download a version of the JDK. In this section, we will use version 8 of the OpenJDK; to install this using apt-get, run "sudo apt-get install openjdk-8-jdk".

If you start a new terminal and run java -version, check that the correct version now shows; if it does not show the correct version, then follow the steps in the previous section.

Downloading and Configuring Apache Spark

Now that you have the correct working version of Java, you can download Apache Spark. Go to the home page (https://spark.apache.org/downloads.html), choose the version you want and the package type, and then download; I downloaded and copied mine to my home directory and then ran

```
$ tar -xvf spark-3.0.0-bin-hadoop2.7.tgz
```

This command extracted the files to ~/ spark-3.0.0-bin-hadoop2.7/, so the next step is to set up a couple of environment variables:

- SPARK_HOME

- PATH

We set SPARK_HOME to the directory that we just extracted, and I would update my .bashrc with the new folder and also update our PATH variable in .bashrc to include "$SPARK_HOME/bin"; make sure you set $SPARK_HOME before you update your path.

Testing the Install

To verify that we have a working installation of Apache Spark, run spark-shell, which is a REPL for running commands. If you run spark-shell, you should see the spark logo displayed and a command prompt where you can run spark commands.

```
$ spark-shell
Setting default log level to "WARN".
To adjust logging level use sc.setLogLevel(newLevel). For SparkR, use
setLogLevel(newLevel).
```

```
Spark context Web UI available at http://localhost:4040
Spark context available as 'sc' (master = local[*], app id =
local-1595401509136).
Spark session available as 'spark'.
Welcome to


      ____              __
     / __/__  ___ _____/ /__
    _\ \/ _ \/ _ `/ __/  '_/
   /___/ .__/\_,_/_/ /_/\_\   version 3.0.0
      /_/

Using Scala version 2.12.10 (OpenJDK 64-Bit Server VM, Java 1.8.0_262)
Type in expressions to have them evaluated.
Type :help for more information.

scala>
```

If you get the scala> prompt, then that is a great sign that everything is working, but let us see if we can run a spark command to do some local processing:

```
scala> spark.sql("select * from range(10)").withColumn("ID2", col("ID")).
show
+---+---+
| id|ID2|
+---+---+
|  0|  0|
|  1|  1|
|  2|  2|
|  3|  3|
|  4|  4|
|  5|  5|
|  6|  6|
|  7|  7|
|  8|  8|
|  9|  9|
+---+---+
```

What we did here was to create a DataFrame by running some spark SQL to call the range function. The range function creates as many rows as we ask it, then we used withColumn to create a second column with the same value that was in the first column. Finally, we used show to display the contents of the DataFrame.

Exiting the spark-shell REPL is like exiting VIM; use :q.

Overriding Default Config

To configure Apache Spark, we can use the configuration files that are in $SPARK_HOME/conf; these are a set of text files that control how Apache Spark works. When you first download Apache Spark, there are only template config files, no actual config files:

```
~/spark-3.0.0-bin-hadoop2.7/conf » ls
docker.properties.template    metrics.properties.template    spark-env.
sh.template
fairscheduler.xml.template    slaves.template
log4j.properties.template    spark-defaults.conf.template
```

If we want to configure any of the config files, we should first copy the template file to the actual file by removing ".template" from the end of the filenames:

```
$ cp ./spark-defaults.conf.template ./spark-defaults.conf
$ cp ./log4j.properties.template ./log4j.properties
```

You can then use your favorite editor to open the config files and edit them. For a development instance of Apache Spark, I would change the log4j section for console logging to just showing errors; showing warnings creates a lot of output that we can normally ignore. If you change the default file contents which looks like:

```
# Set everything to be logged to the console
log4j.rootCategory=INFO, console
log4j.appender.console=org.apache.log4j.ConsoleAppender
log4j.appender.console.target=System.err
log4j.appender.console.layout=org.apache.log4j.PatternLayout
log4j.appender.console.layout.ConversionPattern=%d{yy/MM/dd HH:mm:ss} %p
%c{1}: %m%n
to
```

```
# Set only errors to be logged to the console
log4j.rootCategory=ERROR, console
log4j.appender.console=org.apache.log4j.ConsoleAppender
log4j.appender.console.target=System.err
log4j.appender.console.layout=org.apache.log4j.PatternLayout
log4j.appender.console.layout.ConversionPattern=%d{yy/MM/dd HH:mm:ss} %p
%c{1}: %m%n
```

that is usually a good start. In the spark-defaults.conf file, it is often a good idea to configure how much memory Apache Spark can use on your development machine; my machine, for instance, has 16GB of RAM, so I configure Apache Spark to use 6GB for the executor memory. I also use the Delta Lake library from Databricks quite extensively, so by adding it to my default config, I get access to it every time and do not need to remember to start Apache Spark with the additional library:

```
spark.executor.memory=6g
spark.jars.packages=io.delta:delta-core_2.12:0.7.0
```

Once I have changed the config files, it is always worth rerunning spark-shell to verify that the changes have worked. This time, when I run spark-shell, I can see that the logging is less verbose and also that my delta package has been loaded:

```
$ ./spark-shell
Ivy Default Cache set to: /Users/ed/.ivy2/cache
The jars for the packages stored in: /Users/ed/.ivy2/jars
:: loading settings :: url = jar:file:/Users/ed/spark-3.0.0-bin-without-
   hadoop/jars/ivy-2.4.0.jar!/org/apache/ivy/core/settings/ivysettings.xml
io.delta#delta-core_2.12 added as a dependency
:: resolving dependencies :: org.apache.spark#spark-submit-parent-4987d518-
   30f9-4696-ac0e-1b20ed99f224;1.0
     confs: [default]
     found io.delta#delta-core_2.12;0.7.0 in central
     found org.antlr#antlr4;4.7 in central
     found org.antlr#antlr4-runtime;4.7 in local-m2-cache
     found org.antlr#antlr-runtime;3.5.2 in central
     found org.antlr#ST4;4.0.8 in central
     found org.abego.treelayout#org.abego.treelayout.core;1.0.3 in spark-
     list
```

```
      found org.glassfish#javax.json;1.0.4 in central
      found com.ibm.icu#icu4j;58.2 in central
:: resolution report :: resolve 210ms :: artifacts dl 8ms
      :: modules in use:
      com.ibm.icu#icu4j;58.2 from central in [default]
      io.delta#delta-core_2.12;0.7.0 from central in [default]
      org.abego.treelayout#org.abego.treelayout.core;1.0.3 from spark-list
      in [default]
      org.antlr#ST4;4.0.8 from central in [default]
      org.antlr#antlr-runtime;3.5.2 from central in [default]
      org.antlr#antlr4;4.7 from central in [default]
      org.antlr#antlr4-runtime;4.7 from local-m2-cache in [default]
      org.glassfish#javax.json;1.0.4 from central in [default]
      ---------------------------------------------------------------------
      |                      |            modules      ||    artifacts   |
      |         conf         | number| search|dwnlded|evicted|| number|dwnlded|
      ---------------------------------------------------------------------
      |       default        |   8   |   0   |   0   |   0   ||   8   |   0   |
      ---------------------------------------------------------------------
:: retrieving :: org.apache.spark#spark-submit-parent-4987d518-30f9-4696-
   ac0e-1b20ed99f224
      confs: [default]
      0 artifacts copied, 8 already retrieved (0kB/7ms)
```

Dotnet Worker Configuration

In the introduction, we covered the architecture for .NET for Apache Spark, which is that the .NET code connects to the JVM code using a TCP socket, and so the .NET code can call code in the JVM. There is one exception to this, which is User-Defined Functions or UDFs. UDFs work by running in another separate Microsoft-specific process the JVM connects to and sends messages backward and forward from. If we use UDFs, then we will also need to download another executable from the .NET for the Apache Spark project and configure an environment variable to point to this executable. We will cover more about UDFs in Chapter 4, and you may well never need to use UDFs.

To configure .NET for Apache Spark, so UDFs work, you first need to visit the releases page for the project (`https://github.com/dotnet/spark/releases`), choose the version and operating system you are going to use, download and extract somewhere, and then create the `DOTNET_WORKER_DIR` environment variable to point to the directory with the Microsoft.Spark.Worker executable.

Troubleshooting Common Errors

In this section, we will cover common errors that occur when running Apache Spark.

Unsupported Class File Major Version

When running Apache Spark commands and you see a message "java.lang. IllegalArgumentException: Unsupported class file major version 58", the number at the end is not relevant. This message means that the Java class version is not compatible with the current runtime. This is generally caused by running Apache Spark on an incorrect version of Java. Apache Spark does not validate which version of Java it is running on. Apache Spark attempts to run and then fails if it is the wrong version. Check which version of Java you have configured using "`java -version`", and also in the Apache Spark output, it should print which version of Java it thinks it is using. Until you run Apache Spark and it outputs the correct version of Java for the version of Apache Spark that you have, your programs will always fail.

Exception Exiting Spark

When exiting your Apache Spark application on Windows, you see "ERROR ShutdownHookManager: Exception while deleting Spark temp dir". This is an error you can safely ignore, and if you update your %SPARK_HOME%\conf\log4j.properties with these two lines, then the error messages will be hidden:

```
log4j.logger.org.apache.spark.util.ShutdownHookManager=OFF
log4j.logger.org.apache.spark.SparkEnv=ERROR
```

Cannot Run spark-shell

If you are on Windows and you try to run `spark-shell` but you get an error "The system cannot find the path specified", this can be caused by a semicolon at the end of the JAVA_HOME environment variable or some other error with JAVA_HOME. To ensure your JAVA_HOME variable is correct, try `dir "%JAVA_HOME%\bin\java.exe"`, and if that does not show that java.exe exists, run `echo %JAVA_HOME%\bin\java.exe` and ensure that the entire path points to a java.exe.

If you get another error with Apache Spark, then Stack Overflow is a good place to ask questions, or there is a good community of people who will help if you contact the Apache Spark mailing list. To see all of the available community help, see `https://spark.apache.org/community.html`.

Summary

Getting Apache Spark running on your development machine requires that you install and configure the correct version of Java, which today is Java 8. When the .NET for Apache Spark project supports Apache Spark 3.0, it will be Java 11 with future versions, likely, supporting later versions of Java.

The overall process is to install and configure Java, then download and configure Apache Spark. If you are on Windows, then you should also download winutils.exe. If you wish to use User-Defined Functions, then you will also need the Microsoft .NET worker process, which hosts your UDF code.

If at the end of this chapter you can run spark-shell and you are using the correct version of Java, then you are in good shape and should be excited to write your first .NET for Apache Spark in the next chapter.

## CHAPTER 3

Programming with .NET for Apache Spark

In this chapter, we will be writing our first .NET for Apache Spark application that we can execute and even debug in our favorite .NET IDE. We will cover what it is we need to do in our project and then what we need to do so that we can execute the program using Apache Spark.

Once we have got our first program running, we will move on to look at how to convert from existing PySpark examples and highlight some Scala features that we need to remember when converting from Scala into .NET. The reason for translating from Python and Scala is that they are the most commonly used languages for Apache Spark and will be for quite some time, possibly forever. Understanding how to read the Python and Scala examples and convert the examples to .NET will make you much more effective, and I firmly believe that being effective in .NET for Apache Spark is critical to any implementation.

First Program

In our first program, we will use Apache Spark to create a `DataFrame` and how the data in that `DataFrame`. We will create a `SparkSession`, which is like the gateway to the Apache Spark application, and it is through a `SparkSession` that we get Apache Spark to execute queries. Once we have a `SparkSession`, we will use it to create some data and then do some processing and save the output. Finally, we will take a look at where the data physically sits and how to pull data across from the JVM into our .NET program, including any potential performance issues.

© Ed Elliott 2021
E. Elliott, *Introducing .NET for Apache Spark*, https://doi.org/10.1007/978-1-4842-6992-3_3

I will use the command-line version of dotnet to create a new console application, but please feel free to create a dotnet console project using an IDE.

```
» dotnet new console -output HelloSpark --language "C#"
Getting ready...
The template "Console Application" was created successfully.
Processing post-creation actions...
Running 'dotnet restore' on HelloSpark/HelloSpark.csproj...
  Determining projects to restore...
  Restored /Users/ed/git/scratch/HelloSpark/HelloSpark.csproj (in 121 ms).
Restore succeeded.
```

To create an F# console application, we can use

```
» dotnet new console -output HelloSpark --language "F#"
```

Microsoft.Spark NuGet Package

Once we have a console application, we need to add the Microsoft.Spark NuGet package (www.nuget.org/packages/Microsoft.Spark/).

```
» cd HelloSpark
» dotnet add package Microsoft.Spark
  Determining projects to restore...
  Writing /var/folders/yw/9n3l8f4x2856pxvys69_lh580000gp/T/tmpgLiEtb.tmp
info : Adding PackageReference for package 'Microsoft.Spark' into project
       '/Users/ed/git/scratch/HelloSpark/HelloSpark.csproj'.
info : Restoring packages for /Users/ed/git/scratch/HelloSpark/HelloSpark.
       csproj...
info :   GET https://api.nuget.org/v3-flatcontainer/microsoft.spark/index.
         json
info :   OK https://api.nuget.org/v3-flatcontainer/microsoft.spark/index.
         json 112ms
info : Package 'Microsoft.Spark' is compatible with all the specified
       frameworks in project '/Users/ed/git/scratch/HelloSpark/HelloSpark.
       csproj'.
info : PackageReference for package 'Microsoft.Spark' version '0.12.1'
       added to file '/Users/ed/git/scratch/HelloSpark/HelloSpark.csproj'.
```

```
info : Committing restore...
info : Generating MSBuild file /Users/ed/git/scratch/HelloSpark/obj/
       HelloSpark.csproj.nuget.g.targets.
info : Writing assets file to disk. Path: /Users/ed/git/scratch/HelloSpark/
       obj/project.assets.json
log  : Restored /Users/ed/git/scratch/HelloSpark/HelloSpark.csproj (in 905 ms).
```

These commands have been run on macOS; although the output is slightly different on Windows and Linux, the commands are the same.

If you got this far, then you would have a project that has the Microsoft.Spark NuGet package, and if you followed along in Chapter 1, then you will also have a local Apache Spark instance, which we can now use.

SparkSession

SparkSession is the class that we use to call Apache Spark and get it to do our processing. Each Java Virtual Machine (JVM) can only have a single SparkSession, and there is a specific pattern to getting hold of our SparkSession that involves using SparkSession.Builder() and calling GetOrCreate, which will either get an existing session or create a new session. Listings 3-1 and 3-2 show how to use GetOrCreate to create a SparkSession.

Listing 3-1. Getting a reference to a SparkSession in C#

```
var spark = SparkSession
    .Builder()
    .AppName("DemoApp")
    .Config("some-option", "value")
    .Config("some-other-option", "value")
    .GetOrCreate();
```

Listing 3-2. Getting a reference to a SparkSession in F#

```
let spark = SparkSession
                    .Builder()
                    .AppName("DemoApp")
                    .Config("some-option", "value")
                    .Config("some-other-option", "value")
                    .GetOrCreate()
```

In these code listings, we can see that SparkSession has a static method called Builder(), which returns a Builder. The Builder lets us set various config settings, as well as the application name. There are quite a few config settings we can set, and to see the full set of options, visit http://spark.apache.org/docs/latest/configuration. html#available-properties.

In both of these code listings, I create the SparkSession using the variable name spark. I would suggest that you also use the same name as when using Apache Spark in some environments such as the PySpark or Scala REPL or in Databricks notebooks, the variable spark is defined up front and points to the active SparkSession. If you keep this naming strategy, then it makes it easier to migrate code between environments.

When we use Apache Spark, we need to understand where the code and data reside. If I create a variable and store the string "Hello Apache Spark", the variable and the data will exist in the .NET application; Apache Spark will not be able to see it. Conversely, if I use Apache Spark to read in a file, the data will be available to Apache Spark, but we can't interrogate it in our .NET application. We will either need to write data out somewhere from Apache Spark and read it back from our .NET application, or we will need to Collect() the data back from Apache Spark into our .NET application. Collecting the data back proxies the data back into .NET. If this is a few rows, then it will be fine, but if it is terabytes of data, then this could cause performance issues.

We will now look at our first full example program. In Apache Spark, we are interested in creating and modifying DataFrames, which we can then either aggregate and keep some data or write back out again either to a file or a database. In this next example, Listing 3-3, we will create a DataFrame, save the output to a CSV file, and then also "collect" the results back to .NET so that we can iterate each row and operate on it in .NET.

Listing 3-3. Our first full .NET for Apache Spark application in C#

```csharp
using System;
using Microsoft.Spark.Sql;

namespace ConsoleApp1
{
    class Program
    {
        static void Main(string[] args)
        {
            var spark = SparkSession
                .Builder()
                .AppName("DemoApp")
                .GetOrCreate();

            var dataFrame = spark.Sql("select id, rand() as random_number
            from range(1000)");

            dataFrame
                .Write()
                .Format("csv")
                .Option("header", true)
                .Option("sep", "|")
                .Mode("overwrite")
                .Save(args[1]);

            foreach (var row in dataFrame.Collect())
            {
                if (row[0] as int? % 2 == 0)
                {
                    Console.WriteLine($"row: {row[0]}");
                }
            }
        }
    }
}
```

If we break this down, in Listing 3-4, we start with our SparkSession, and we create that by using SparkSession.Builder(), which in turn we get from SparkSession.

Listing 3-4. Creating a SparkSession in C#

```
var spark = SparkSession
    .Builder()
    .AppName("DemoApp")
    .GetOrCreate();
```

When we create the SparkSession, we can, optionally, pass in an AppName. This is useful when we are troubleshooting a shared Apache Spark instance as the AppName is displayed in the SparkUI which we look at in detail in Chapter 10.

We then show in Listing 3-5 that we can create a DataFrame using spark.Sql and passing in a SQL query. In these demos, I use range to create rows; in a real-life project, you would more likely create DataFrames by reading in data.

Listing 3-5. Using SQL to create 1000 rows in C#

```
var dataFrame = spark.Sql("select id, rand() as random_number from
range(1000)");
```

Here, we pass two functions to spark.Sql. We pass rand(), which creates a random number for every row and range, which creates rows, in this case, rows with an ID column of between 0 and 999. In Listing 3-6, we show how to write a DataFrame to disk. The things to note here are that we specifically pass in "csv" as the format. However, the default CSV options are rarely ideal, so we can override the defaults to force the column headers to be written and also to change the default separator using the options "header" and "sep"; for a full list of CSV format options, see the csv method in the DataWriter API documentation: https://spark.apache.org/docs/latest/api/java/org/apache/spark/sql/DataFrameWriter.html.

Listing 3-6. Writing the DataFrame to disk using C#

```
dataFrame
        .Write()
        .Format("csv")
        .Option("header", true)
```

```
    .Option("sep", "|")
    .Mode("overwrite")
    .Save(args[1]);
```

At this point, the data is only on the JVM side, and we can ask Apache Spark to write the data or perform any operation we want on the data, but we can't see the data in .NET. We can see that there is a DataFrame and count the rows, but to bring the data back over to .NET, we need to call Collect(). In Listing 3-7, we show Collect, which returns an IEnumerable<Row>, which lets us access each row and each column on each row.

Listing 3-7. Writing the DataFrame to disk using C#

```
foreach (var row in dataFrame.Collect())
{
  if (row[0] as int? % 2 == 0)
  {
    Console.WriteLine($"row: {row[0]}");
  }
}
```

Now we have our full program in C#; we will also show the same program, this time, implemented in F# in Listing 3-8.

Listing 3-8. Our first full .NET for Apache Spark application in F#

```
[<EntryPoint>]
let main argv =

    let spark = SparkSession
                    .Builder()
                    .AppName("DemoApp")
                    .GetOrCreate()

    let dataFrame = spark.Sql("select id, rand() as random_number from
    range(1000)")

    dataFrame
            .Write()
            .Format("csv")
```

```
        .Option("header", true)
        .Option("sep", "|")
        .Mode("overwrite")
        .Save(argv.[1]);

dataFrame.Collect()
            |> Seq.map(fun row -> row.Get(0) :?> int)
            |> Seq.filter(fun id -> id % 2 = 0)
            |> Seq.iter(fun i -> printfn "row: %d" i)

0
```

The F# version is pretty much the same as the C# version until we want to collect the data back to .NET where we map the IEnumerable<Row> to retrieve the specific column we want, then filter and iterate the rows.

In the next section, we will cover how to run our program on our local Apache Spark instance and then how we can debug the .NET code in our favorite IDE.

Executing the Program

In Chapter 1, we covered what was involved in writing .NET code that ended up calling Java classes and methods inside a Java Virtual Machine (JVM). To summarize, what we need to do is to start a session of Apache Spark and use the Java class supplied by the .NET for Apache Spark project that starts our .NET process and proxies requests between the .NET code and the JVM code. In this section, we will cover how to run our applications and then how to debug the code in our favorite IDE.

Executing on the Command Line

To execute our program, we need to run spark-submit, which starts the Java Virtual Machine and initializes an instance of Apache Spark. We pass a parameter to spark-submit which tells Apache Spark to load the JAR file that is shipped with the Microsoft. Spark NuGet package and run the org.apache.spark.deploy.dotnet.DotnetRunner class. The DotnetRunner class starts a listening port, then runs our application, and when our application uses the SparkSession.Builder() to connect to Apache Spark, the DotnetRunner then accepts the connection and passes requests and responses between the two processes.

The full command line is, therefore, the spark-submit script, the class name we want to instantiate, the path to the Microsoft.Spark JAR file, the command to execute *our .NET program,* and any arguments that our program requires. We show these arguments in Table 3-1.

Table 3-1. *Minimum arguments to run our .NET for Apache Spark application*

Argument	Notes	Example
spark-submit	Starts Apache Spark	spark-submit
--class	The full class name of the DotnetRunner	org.apache.spark.deploy. dotnet.DotnetRunner
Path to JAR file	The JAR is in the NuGet package and copied to the bin folder on build I believe it is best to be explicit and pass the full path because Apache Sparks working directory will likely not be your working directory On Mac/Linux, this will be case sensitive, and on Windows if you have spaces anywhere in the path, you will need to use quotes to wrap the path, so avoid spaces where possible	/Users/user/git/HelloSpark/ bin/Debug/netcoreapp3.1/ microsoft-spark-2.4.x- 0.12.1.jar
Our program and any arguments		dotnet run --project / project/path /tmp/csv- output

If you download the sample code for this book and extract the files, build the solution, and then you will find the JAR file that will have been copied to the bin directory.

The JAR is specific to each version of Apache Spark, and the version of both Apache Spark and the Microsoft.Spark library is encoded in the filename. For example, the JAR file "microsoft-spark-2.4.x-0.12.1.jar" is for Apache Spark version "2.4.x" and is for the Microsoft.Spark library version "0.12.1".

If you are not sure which version of Apache Spark you are using, you can run "spark-shell --version", and the version will be displayed on the screen. If, for example, after you build the solution and your JAR file path is "c:\code\dotnet-spark\ch03\ Chapter03\Listing3-3\bin\Debug\netcoreapp3.1\ microsoft-spark-2.4.x-0.12.1.jar" and Listing3-3 requires a single argument which is a path to write the csv file to, then your command line would be

```
spark-submit --class org.apache.spark.deploy.dotnet.DotnetRunner "c:\code\
dotnet-spark\ch03\Chapter03\Listing3-3\bin\Debug\netcoreapp3.1\ microsoft-
spark-2.4.x-0.12.1.jar" dotnet run "c:\code\dotnet-spark\ch03\Chapter03\
Listing3-3" "c:\code\dotnet-spark\ch03\Chapter03\output-csv"
```

When we run, we should see the following text among the output:

```
[2020-08-13T09:34:00.5421480Z] [Machine] [Info] [ConfigurationService]
Using port 49596 for connection.
[2020-08-13T09:34:00.5623630Z] [Machine] [Info] [JvmBridge] JvMBridge port
is 49596
line: 0
line: 2
line: 4
line: 6
```

Further, if we look in the output directory that we passed in, we should see a set of CSV files. What often confuses new users of Apache Spark is that although you passed in a path to a CSV file, a directory was created, and inside the directory, you get 12 CSV files. There are 12 files because Apache Spark splits its processing among its executors, and 12 different executors wrote the files. The benefit of splitting a large file into smaller files is when Apache Spark reads the files back in for processing later, then each file can be read independently which helps to easily parallelize the load. When you run in production, you can tell the performance difference from reading several individual files compared to reading from a compressed file, which, due to the nature of compression, means that only a single executor can read the file at a time.

It is possible to get Apache Spark to write a single file or any number of files you determine using Coalesce or Repartition on the DataFrame before writing the DataFrame out, but doing so may cause performance issues later on.

Debugging the .NET Code in an IDE

Running our programs is great, but sometimes we want to debug our code, and in .NET, that means using a debugger such as Visual Studio, Visual Studio Code, or JetBrains Rider. When we run in a debugger, we still need to account for the fact that we need to start the Apache Spark instance on a JVM that is a separate process than our .NET application.

To debug our applications, we can use the debug argument which we can pass to the Scala DotnetRunner class. Instead of running our program, the DotnetRunner will create the listening port and wait for an incoming connection. Using the debug command allows us to, separately, start a debug session of our application and connect to the Apache Spark instance.

The command line is similar to the last one we used. The difference is that we will pass the word "debug" instead of the command to run our .NET program:

```
spark-submit --class org.apache.spark.deploy.dotnet.DotnetRunner "c:\code\
dotnet-spark\ch03\Chapter03\Listing3-3\bin\Debug\netcoreapp3.1\ microsoft-
spark-2.4.x-0.12.1.jar" debug
```

When we run this, we should see the following output:

```
************************************************************
* .NET Backend running debug mode. Press enter to exit *
************************************************************
```

Then we can start our application in our favorite debugger, and when we call SparkSession.Builder(), our application connects to the already running instance of Apache Spark.

When we run like this, we can set breakpoints and examine local variables as you would expect in a .NET application. It is important to remember that although we see objects like a SparkSession and a DataFrame, these are not the actual SparkSession or DataFrame that exists in the JVM but a reference to the real objects, so we can't see the data that is inside them, unless we use Collect() to bring the data across. In Figure 3-1, we see the JetBrains Rider IDE, which has a breakpoint set and is displaying information about the local variables.

```
C# Program.cs ×

14
15          var dataFrame = spark.Sql("select id, rand() as random_number from range(1000)");  dataFrame: {Microsoft.Spark.Sql.DataFrame}
16
17          dataFrame  dataFrame: {Microsoft.Spark.Sql.DataFrame}
18              .Write()
19              .Format("csv")
20              .Option("header", true)
21              .Option("sep", "|")
22              .Mode("overwrite") // DataFrameWriter
23              .Save(args[0]);  args: Count = 1
24
25          foreach (var row in dataFrame.Collect())  row: "[0,0.46249144700872]"  dataFrame: {Microsoft.Spark.Sql.DataFrame}
26          {
27              if (row[0] as int? % 2 == 0)  row: "[0,0.46249144700872]"
28              {
29                  Console.WriteLine($"line: {row[0]}");       |          row: "[0,0.46249144700872]"
30              }
31          }
32          }
33      }
34  }
```

Figure 3-1. *Debugging our .NET for Apache Spark application*

With a debugger like JetBrains Rider or Visual Studio, we can step into and over our code, set breakpoints, and inspect local variables. Figure 3-2 shows the contents of the variables. The objects that are references to JVM objects such as `dataFrame` or `spark` are references, and we can't see any useful properties. Variables, such as `row`, which have been collected back to the .NET application, are available in .NET. In this case, there are 1000 rows, so do not take up much memory or cost very much to proxy from the JVM, but if there were billions of rows, then this would likely be an issue.

Figure 3-2. *Exploring local variables*

The exciting thing to see is that although the dataFrame variable is a reference to the JVM object and we can't see any of the actual properties, we can still evaluate expressions. In Figure 3-3, I used the evaluation window in Rider to evaluate dataFrame. Count(), and the result was displayed. In Visual Studio, you could use the immediate window or the watch window to do something similar.

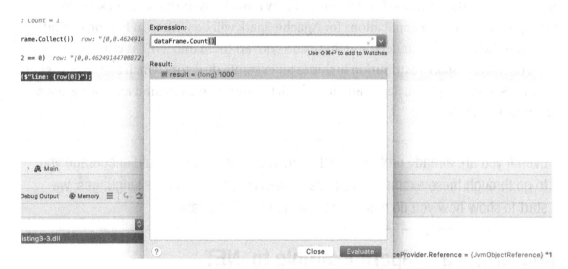

Figure 3-3. *Examining local variables*

Further Use of the Debug Command

Usually, when your application exits, the Apache Spark instance and Java Virtual Machine also exit. When you run the debug command, the Apache Spark instance stays alive, and so you can start a new instance of your application and connect to the previous instance of Apache Spark. Keeping a running instance of Apache Spark is useful in a couple of cases. The first is, when running unit and integration tests, if you require an Apache Spark instance, you can start a single instance and reuse it for all of your tests, which keeps the overhead of starting Apache Spark to a minimum. The second case is where you are developing an application, if you have a debug instance running in the background, then you can quickly test changes without having to start and stop an Apache Spark instance every time.

When the debug instance starts, it listens on a specific port, which is 5567, and you can start and use multiple debug instances by changing the default port using the environment variable DOTNETBACKEND_PORT. If you change the port that the debug instance listens on, then you also need the environment variable in your .NET process.

Converting Existing Applications to .NET

In this section, we will look at how to read PySpark and Scala code that have been written for Apache Spark and how we can convert these into .NET code. There are a few things to understand, and even if we do not know how to program in Python or Scala, we should be able to read the code and have an idea of how it works as the majority of samples, examples, and existing applications for Apache Spark will be written in Python or Scala; therefore, being able to understand these and convert them into .NET is critical.

The Apache Spark distribution includes an example directory which contains a set of examples in Java, Python, Scala, and R. We will be converting a Python and also a Scala example into .NET.

Even if you are already proficient with Python and/or Scala, I would encourage you to go through these examples because as well as introducing the languages, we start to show how you do basic processing in Apache Spark.

Converting a PySpark Example to .NET

In the Apache Spark distribution example directory, we will convert the "src/main/python/sql/basic.py" example. The file used in this book is also included in the sample code as Listing 3-9. The C# version of the code is in the included project called "PythonToCSharp", and the F# version is called "PythonToFSharp".

In Listing 3-9, we see what is called the docstring in Python, and this is the way Python documents code and is similar to the XML comments we often add to C# methods and classes.

Listing 3-9. Python docstring or documentation

```
"""
A simple example demonstrating basic Spark SQL features.
Run with:
   ./bin/spark-submit examples/src/main/python/sql/basic.py
"""
```

Python is very hierarchical in that you can import other modules at the module level, the file level, the class level, or the function level, and wherever an import occurs, the import is available at levels below, so when you need to see what is imported, you must start at the function and check upward for more imports. In Listing 3-10, we can see four imports, and, aside from the last import, "from pyspark.sql.types import *", which imports everything under "pyspark.sql.types", each statement imports one function or class from a module. Another way the imports could have been written is either "import pyspark.sql" or "from pyspark.sql import SparkSession, Row".

Listing 3-10. Python imports

```
from __future__ import print_function

# $example on:init_session$
from pyspark.sql import SparkSession
# $example off:init_session$

# $example on:schema_inferring$
from pyspark.sql import Row
# $example off:schema_inferring$

# $example on:programmatic_schema$
# Import data types
from pyspark.sql.types import *
# $example off:programmatic_schema$
```

In Listings 3-11, C#, and 3-12, F#, we show instead of importing pyspark.XXX, we change the reference to Microsoft.Spark.XXX, so pyspark.sql.SparkSession becomes Microsoft.Spark.Sql.SparkSession.

Listing 3-11. Python imports converted to C#

```
using Microsoft.Spark.Sql;
using Microsoft.Spark.Sql.Types;
```

Listing 3-12. Python imports converted to F#

```
open Microsoft.Spark.Sql
open Microsoft.Spark.Sql.Types
```

In Listing 3-13, we see our first function defined. In Python, the structure of a program is defined using whitespace, so where we see a "def name():", everything that is indented, at least one level, further down the file is part of the same function. This function is called "basic_df_example" and includes a single parameter called "spark". The first thing the function does is to read in a JSON file using "spark.read.json". The contents of the JSON file are saved to a DataFrame and then displayed using show().

Listing 3-13. Defining a function in Python and reading in a JSON file as a DataFrame

```
def basic_df_example(spark):
    df = spark.read.json("examples/src/main/resources/people.json")
    # Displays the content of the DataFrame to stdout
    df.show()
```

In Listing 3-14, C#, we will create a method that calls the .NET version. In C#, we need to specify the type of the "spark" parameter, which will be SparkSession. We will see where spark is defined later in the file. In Listing 3-15, F#, we will create a function that also reads the JSON file and displays the contents.

Listing 3-14. Defining a method in C# and reading in a JSON file as a DataFrame

```
static void BasicDfExample(SparkSession spark)
{
    var dataFrame = spark.Read().Json("examples/src/main/resources/people.
    json");
    dataFrame.Show();
}
```

Listing 3-15. Defining a function in F# and reading in a JSON file as a DataFrame

```
    let BasicDfExample (spark:SparkSession) =
        let dataFrame = spark.Read().Json("examples/src/main/resources/
        people.json")
        dataFrame.Show()
```

The Python code then prints the schema of the DataFrame in Listing 3-16, which is a handy way to, visually, see whether you have read a file correctly.

Listing 3-16. Printing the schema of a DataFrame

```
df.printSchema()
```

In Listings 3-17 and 3-18, we need to convert `printSchema` into the .NET equivalent, which is `PrintSchema`. When you switch from Python to .NET, changing the capitalization of the methods is the most common thing to do.

Listing 3-17. Printing the schema of a DataFrame in C#

```
dataFrame.PrintSchema()
```

Listing 3-18. Printing the schema of a DataFrame in F#

```
dataFrame.PrintSchema()
```

In Listing 3-19, the Python code is selecting a single column, by name, and then printing out the resulting `DataFrame`. Calling `select` on a `DataFrame` creates a new instance of the `DataFrame`, so any operations do not affect other parts of your code.

Listing 3-19. Selecting a single column and displaying the contents

```
df.select("name").show()
```

In Listings 3-20 and 3-21, we show the C# and F# versions.

Listing 3-20. Selecting a single column and displaying the resulting DataFrame in C#

```
dataFrame.Select("name").Show();
```

Listing 3-21. Selecting a single column and displaying the resulting DataFrame in F#

```
dataFrame.Select("name").Show()
```

In Listing 3-22, the code selects two columns instead of using the name of the column, as we saw in Listing 3-19. What we see here is the DataFrame being indexed with the name of the columns, and what you get is the Column object. The Column object is important as it defines the name of the column and also the source DataFrame. Consider if you had two DataFrames and they both included the column "ID", if you joined the two DataFrames and did a Select("ID"), which column would you get? It isn't possible to tell, which would cause Apache Spark to fail. In this case, you can specify exactly which ID column you want by retrieving the Column reference from the DataFrame rather than using the string name. The second thing that we see is that the age column is a calculation, and it is important to understand that the calculation is passed into Apache Spark, so the Apache Spark runtime runs the calculation, rather than the data being passed to the Python process and Python running the calculation. For large datasets, this really matters for performance.

Listing 3-22. Selecting two columns and performing a calculation on the age column

```
df.select(df['name'], df['age'] + 1).show()
```

In Listings 3-23, in C#, and 3-24, F#, we can do the same thing, but should note that when we call dataFrame["age"] +1, it is *syntactic sugar* for dataFrame["age"].Plus(1), and it may well be clearer to write it as such.

Listing 3-23. Selecting multiple columns and performing a calculation in C# including the more explicit ".Plus" call

```
dataFrame.Select(dataFrame["age"], dataFrame["age"] + 1).Show();
dataFrame.Select(dataFrame["age"], dataFrame["age"].Plus(1)).Show();
```

Listing 3-24. Selecting multiple columns and performing a calculation in F# including the more explicit ".Plus" call

```
dataFrame.Select(dataFrame.["name"], (dataFrame.["age"] + 1)).Show()
dataFrame.Select(dataFrame.["name"], (dataFrame.["age"].Plus(1))).Show()
```

In Listing 3-25, the code filters a DataFrame, and again this is using syntactic sugar to hide the call to the Gt function of the Column object.

Listing 3-25. Filtering a DataFrame using the Gt function on the Column object

```
df.filter(df['age'] > 21).show()
```

Listings 3-26 and 3-27 show the .NET version of the filters, and, again, it is essential to understand that the actual calculation happens on the JVM side inside Apache Spark rather than running in the .NET code.

Listing 3-26. Filtering a DataFrame using the Gt function on the Column object, in C#

```
dataFrame.Filter(dataFrame["age"].Gt(21)).Show();
```

Listing 3-27. Filtering a DataFrame using the Gt function on the Column object, in F#

```
dataFrame.Filter(dataFrame["age"].Gt(21)).Show();
```

In Listing 3-28, we can see an aggregation on the DataFrame by grouping on age and counting how many rows of each age there are; the C# and F# versions are in Listings 3-29 and 3-30.

Listing 3-28. Aggregations in Apache Spark

```
df.groupBy("age").count().show()
```

Listing 3-29. Aggregating DataFrames in C#

```
dataFrame.GroupBy(dataFrame["age"]).Count().Show();
```

Listing 3-30. Aggregating DataFrames in F#

```
dataFrame.GroupBy(dataFrame.["age"]).Count().Show()
```

In Listing 3-31, we see that the DataFrame is converted into an Apache Hive view, which allows the DataFrame to be queried using SQL statements. The SparkSession class has the Sql() method, which allows us to run SQL statements. However, there is no way in that SQL context to see data in a DataFrame, *unless* we take the DataFrame and make it available to the SQL context, which we do using a temp view. A temp view will live as long as the SparkSession, so when the session ends, so does the temp view.

Listing 3-31. Making a DataFrame accessible by the SQL context

```
df.createOrReplaceTempView("people")
sqlDF = spark.sql("SELECT * FROM people")
sqlDF.show()
```

Listing 3-32 shows how to create the view in C# and Listing 3-33 which is in F#

Listing 3-32. Making a DataFrame accessible by the SQL context in C#

```
dataFrame.CreateOrReplaceTempView("people");
var sqlDataFrame = spark.Sql("SELECT * FROM people");
```

Listing 3-33. Making a DataFrame accessible by the SQL context in F#

```
dataFrame.CreateOrReplaceTempView("people")
let sqlDataFrame = spark.Sql("SELECT * FROM people")
```

There are some variations on the CreateOrReplaceTempView function. You can use CreateTempView, which will not overwrite an existing view, or you can use CreateGlobalTempView or CreateOrReplaceGlobalTempView. A global temp view is available to other SparkSessions in the running Apache Spark instance even after your session stops, but temp views are destroyed when your session ends. In Listing 3-34, we see a global temp view created, which incidentally means we need to prefix the name of the view with "global_temp" when we want to use it in the SQL context.

Listing 3-34. Making a DataFrame accessible to other SparkSession's SQL context

```
df.createGlobalTempView("people")
spark.sql("SELECT * FROM global_temp.people").show()
spark.newSession().sql("SELECT * FROM global_temp.people").show()
```

In Listing 3-35, we will see how to make a DataFrame accessible to other SparkSession's in C#.

Listing 3-35. Making a DataFrame accessible to other SparkSession's SQL context in C#

```
dataFrame.CreateGlobalTempView("people");
spark.Sql("SELECT * FROM global_temp.people").Show();
spark.NewSession().Sql("SELECT * FROM global_temp.people").Show();
```

In Listing 3-36, we will see how to make a DataFrame accessible to other SparkSession's in F#.

Listing 3-36. Making a DataFrame accessible to other SparkSession's SQL context F#

```
dataFrame.CreateGlobalTempView("people")
spark.Sql("SELECT * FROM global_temp.people").Show()
spark.NewSession().Sql("SELECT * FROM global_temp.people").Show()
```

The last bit of Python code that we will look at in this section is the standard main definition of a Python script file in Listing 3-37.

Listing 3-37. Python script entry point

```python
if __name__ == "__main__":
    # $example on:init_session$
    spark = SparkSession \
        .builder \
        .appName("Python Spark SQL basic example") \
        .config("spark.some.config.option", "some-value") \
        .getOrCreate()
    # $example off:init_session$

    basic_df_example(spark)
```

This is standard in Python scripts. What it means is that if you run the script intentionally, then the code inside this `if` statement is run. However, if you run a separate file that imports this Python file, then the code will not be executed.

In Listing 3-37, we also see that `SparkSession builder` is used to `getOrCreate` a SparkSession, which is passed to the `basic_df_example` function. Listings 3-38 and 3-39 show the C# and F# equivalents.

Listing 3-38. C# entry point

```csharp
static void Main(string[] args)
{
    var spark = SparkSession.Builder().GetOrCreate();
    BasicDfExample(spark);
}
```

Listing 3-39. F# entry point

```
let spark = SparkSession.Builder().GetOrCreate()
BasicDfExample spark
```

Converting Scala Examples to .NET

When we look at a Scala Apache Spark application, there are two types of code we see; firstly, we see the Scala code that does things like process command-line arguments, and secondly, we see the Scala code that calls Apache Spark classes and methods. Many of the calls to Apache Spark are the same in either Python or Scala and similar in .NET except for the different naming standards. Listing 3-40 shows a valid Python or a valid Scala statement.

Listing 3-40. Valid Python or valid Scala

```
df.printSchema()
```

Instead of walking through line by line as we did for the PySpark version of the code, there are some things in Scala that you often come across in Apache Spark examples that it will be helpful to understand.

Referencing Columns Using $

In Scala, there is a shortcut to getting a reference to a Column. If you are querying a DataFrame and you need to supply a column, you can write $"ColumnName", so if you see something like Listing 3-41, then you can see that Scala can use that shortcut rather than the slightly longer version PySpark or .NET needs, as in Listing 3-42.

Listing 3-41. Scala shortcut to reference a column

```
dataFrame.Select($"ColumnName")
```

Listing 3-42. .NET has no shortcut to reference a column

```
dataFrame.Select(dataFrame["ColumnName"])
```

Datasets

Scala has a feature that is not available in PySpark or .NET, which is statically typed Datasets. Statically typed Datasets allow you to read from a data source, and each row is validated against a known class, so when you run, you know that every row conforms to the correct type. This is a really compelling use case for Scala, but we don't have it available today in .NET.

Where you see a `DataFrame` being read in and then converted to a `Dataset` using `.as[Type]`, beware that the code could reference the columns by the property name rather than passing in the name of the column; for an example of this, see Listing 3-43, where a filter uses a lambda function to reference age as a standard class property.

Listing 3-43. Scala referencing a column as a Dataset

```
case class Person(
    name: String,
    age: Long
)

val people = dataFrame.as[Person]
people.filter(p => p.age > 10)
```

If you see Datasets being used in an example, then you will need to use the `Column` references we have in .NET.

Summary

In this chapter, we have written our first .NET for Apache Spark application, run the program on the command line, and debugged the application. We have also taken a look at how to convert PySpark example code into .NET example code and highlighted a couple of "gotchas" when reading Scala Apache Spark code we need to bear in mind. Hopefully, you have been able to follow along and get your application running on your developer machine.

In the next chapter, we will be looking at what a User-Defined Function is in Apache Spark and how and where we used them as well as how to debug them as they add a layer of complexity over the complexity of the multiple processes we already have.

PART II

The APIs

CHAPTER 4

User-Defined Functions

When we execute Apache Spark code from .NET, we are calling methods and classes in the Java Virtual Machine, and Apache Spark reads, writes, aggregates, and transforms our data, according to our requirements. It is entirely possible and quite common that the .NET application never has any sight of the actual data, and the JVM handles all of the data modifications. This is fine if Apache Spark has all of the classes and methods you need to complete your processing. However, what do we do when we need to do something that isn't supported by Apache Spark? The answer is User-Defined Functions (UDFs) and User-Defined Aggregate Functions (UDAFs). UDFs and UDAFs allow us to bring the data back into .NET and then run any processing that we can dream of in .NET.

An Example

We will visit User-Defined Aggregate Functions later on; instead of operating on each row of a DataFrame, UDAFs work with aggregated data. If you wanted to implement your own Group By, Sum, or Count without using the native Apache Spark code, then you would write a UDAF.

In Listing 4-1, we can see where we have .NET code that is operating on each row and returning a new value, we pass in a column, and we get a new column back.

Listing 4-1. Calling a .NET UDF from Apache Spark in C#

```
var spark = SparkSession.Builder().GetOrCreate();

Func<Column, Column> udfIntToString = Udf<int, string>(id => IntToStr(id));

var dataFrame = spark.Sql("SELECT ID from range(1000)");

dataFrame.Select(udfIntToString(dataFrame["ID"])).Show();
```

67

© Ed Elliott 2021
E. Elliott, *Introducing .NET for Apache Spark*, https://doi.org/10.1007/978-1-4842-6992-3_4

```
string IntToStr(int id)
{
    return $"The id is {id}";
}
```

Listing 4-2. Calling a .NET UDF from Apache Spark in F#

```
let main argv =

    let spark = SparkSession.Builder().GetOrCreate()
    let udfIntToString = Microsoft.Spark.Sql.Functions.Udf<int, string>
    (fun (id) -> "The id is " + id.ToString())
    let dataFrame = spark.Sql("SELECT ID from range(1000)")
    dataFrame.Select(udfIntToString.Invoke(dataFrame.["ID"])).Show()
    0
```

The output from both Listings 4-1 and 4-2 is as follows:

```
+------------------------------------+
|System.String <Main>b__0_0(Int32)(ID)|
+------------------------------------+
|                         The id is 0|
|                         The id is 1|
|                         The id is 2|
|                         The id is 3|
|                         The id is 4|
|                         The id is 5|
|                         The id is 6|
|                         The id is 7|
|                         The id is 8|
|                         The id is 9|
|                        The id is 10|
|                        The id is 11|
|                        The id is 12|
|                        The id is 13|
|                        The id is 14|
|                        The id is 15|
|                        The id is 16|
```

```
|                                    The id is 17|
|                                    The id is 18|
|                                    The id is 19|
+------------------------------------+
only showing top 20 rows
```

Architecture

The way .NET UDFs work is similar to the way that the .NET driver works in that the Java DotNetRunner class launches the .NET code, and a network socket is opened and used to proxy requests backward and forward. Apache Spark creates the UDF process, which it happens to think is a Python process. Apache Spark waits for a socket to be created, and then Apache Spark sends the data down and expects some data in response. Apache Spark has had this functionality for quite some time and is the same process that is used for Python and R support. If you remember that the .NET for Apache Spark process works because the Apache Spark JVM process starts the .NET process, we now have the JVM, the .NET process, and now a second .NET worker process which is used to receive the data for the UDF and to call the .NET code.

· There is some complexity to understand here because the .NET worker process loads the .NET application as a library and using reflection finds the code that is to be executed as the UDF. This means that although we have our .NET process, any UDF work is executed in a separate process, so any initialization outside of the UDF may not have taken place. Any shared state will be lost, so keep processing that you need within the UDF. In Listing 4-3, we will look at a UDF and how any shared state is lost. In Listing 4-4, we look at how in F# we need to explicitly mark a variable as mutable which isn't generally recommended in F#, and even then the shared state is lost; ironically, the easiest way to write UDFs in F# is to follow the guidance of not having shared state.

Listing 4-3. Shared state is lost when running UDFs in C#

```
private static int AddAmount = 100;

private static Int64DataFrameColumn Add100(Int64DataFrameColumn id)
{
    return id.Add(AddAmount);
}
```

```
static void Main(string[] args)
{
    var spark = SparkSession.Builder().GetOrCreate();

    var d = spark.Sql("SELECT ID FROM range(1000)");

    AddAmount = 991923;

    var addUdf = VectorUdf<Int64DataFrameColumn, Int64DataFrameColumn>((id)
    => Add100(id));
    d.Select(d["ID"], addUdf(d["ID"])).Show();
}
```

Listing 4-4. Explicitly creating a mutable shared variable in F#, the altered, changed state is not available in the UDF process

```
let mutable SharedState = 100

[<EntryPoint>]
let main argv =

    let spark = SparkSession.Builder().GetOrCreate();

    let dataFrame = spark.Sql("SELECT ID FROM range(1000)")
    SharedState = 991923
    let addUdf =  Microsoft.Spark.Sql. DataFrameFunctions.
    VectorUdf<Int64DataFrameColumn, Int64DataFrameColumn>(fun (column) ->
    column.Add(SharedState));
    dataFrame.Select(dataFrame.["ID"], addUdf.Invoke(dataFrame.["ID"])).
    Show();

    0
```

The output from this shows that even though in the main .NET process we set the variable AddAmount to 991923, the new column uses the value that AddAmount was initialized to, 100.

```
» spark-submit --class org.apache.spark.deploy.dotnet.DotnetRunner
./microsoft-spark-2.4.x-0.12.1.jar dotnet ./Listing4-3.dll
```

```
+---+------------------------------+
| ID|Int32 <Main>b__0_1(Int32)(ID)|
+---+------------------------------+
|  0|                          100|
|  1|                          101|
|  2|                          102|
|  3|                          103|
|  4|                          104|
|  5|                          105|
|  6|                          106|
|  7|                          107|
|  8|                          108|
|  9|                          109|
| 10|                          110|
| 11|                          111|
| 12|                          112|
| 13|                          113|
| 14|                          114|
| 15|                          115|
| 16|                          116|
| 17|                          117|
| 18|                          118|
| 19|                          119|
+---+------------------------------+
only showing top 20 rows
```

There are a couple of important parts about paths. The first is that the worker process needs to be downloaded from https://github.com/dotnet/spark/releases and placed in a directory. There should then be an environment variable called "DOTNET_WORKER_PROCESS", which points to the directory. Inside the directory, there should be an executable called Microsoft.Spark.Worker.

When we run our .NET application without any UDFs, then we can use "dotnet run --project" and pass either the project csproj/fsproj file or the directory with the project file in, and the application will start. With UDF code, there is an assembly loader, which means you either need to call "dotnet run Listing4-3.dll" and the full path to the dll or be in the same directory as the dll. It can be quite tricky to get the correct paths working for UDFs, but if you have a directory with your compiled application and you run the application from there, then the UDF assembly loader should find the correct code. If the assembly loader cannot find the dll to load when running the UDF, you will see an error message with a `System.IO.FileNotFoundException` exception.

If you do not have the Microsoft.Spark.Worker application or the DOTNET_ WORKER_PROCESS configured so that Apache Spark can find the process and can execute the program, then you will get errors, and the UDF will not run.

Performance

Apache Spark is optimized for performance. The file formats that it uses, such as Parquet, are optimized for performance, typically with columnar data formats that help process large data efficiently, skipping columns that are not needed for whatever the current process happens to be. Passing data between processes adds a performance overhead, and this cannot be avoided. The original UDF support in Apache Spark used what is called Python Pickling, which is a way to serialize and deserialize the data that is sent over the connection. Python Pickling is quite an expensive way to send data and works at the row level, so if you had a UDF that read one column from a dataset of hundreds of columns, each column was pickled and sent down the connection. The difference in performance between applications that kept data on the JVM side and applications that pickled data to another process was quite significant.

Apache Arrow was created to make it much more efficient to share data between processes. Instead of Pickling, Apache Arrow is a columnar format, so only the columns used by the UDF are transferred between the processes.

The general recommendation is always to avoid UDFs for performance reasons if you are not writing in Scala or Java, but if you need to use them, and you care about performance, then you do have some options. One workaround that people used to do is to write UDFs in Scala or Java and register them but use PySpark to call the UDF. Listing 4-5 shows how to do this in C#, and Listing 4-6 shows how to do this in F#. Note that we need to have a Java class that is inside a JAR file that has been added to the "classpath" of the Apache Spark instance for these next two samples to execute.

Listing 4-5. Registering a Java UDF and calling that from Spark SQL and from the DataFrame API in C#

```
var spark = SparkSession.Builder().GetOrCreate();

spark.Udf().RegisterJava("java_function", "com.company.ClassName");

var dataFrame = spark.Sql("SELECT ID, java_function(ID) as java_function_
output FROM range(1000)");
dataFrame.Select(CallUDF("java_udf", dataFrame["ID"])).Show();
```

Listing 4-6. Registering a Java UDF and calling that from Spark SQL and from the DataFrame API in F#

```
    let spark = SparkSession.Builder().GetOrCreate()
    spark.Udf().RegisterJava("java_function", "com.company.ClassName")
    let dataFrame = spark.Sql("SELECT ID, java_function(ID) as java_
    function_output FROM range(1000)")
    dataFrame.Select(Microsoft.Spark.Sql.Functions.CallUDF("java_udf",
    dataFrame.["ID"])).Show();
```

Pickling

If we are content with not having to worry about the performance of our .NET for Apache Spark applications and we have to use a UDF to achieve our goals, we can use pickling to call a UDF. Listings 4-7 and 4-8 show how to use pickling in C# and F#, respectively. We define a function using native types and call that directly. Even though there is nothing that explicitly calls this out in the code as using the old-style pickling, we need to be aware that any data will be pickled and will likely be slow for large datasets.

Listing 4-7. A pickling UDF in C#

```
static void Main(string[] args)
{
    var spark = SparkSession.Builder().GetOrCreate();

    var dataFrame = spark.Sql("SELECT ID FROM range(1000)");
```

```
    var add100 = Udf<int?, int>((input) => input + 100 ?? 100);

    dataFrame.Select(add100(dataFrame["ID"])).Show();
}
```

Listing 4-8. A pickling UDF in F#

```
let spark = SparkSession.Builder().GetOrCreate()
    let dataFrame = spark.Sql("SELECT ID FROM range(1000)")
    let add100 = Functions.Udf<System.Nullable<int>, int>(fun input -> if
input.HasValue then input.Value + 100 else 100 )
    dataFrame.Select(add100.Invoke(dataFrame.["ID"])).Show()
    0
```

Apache Arrow

To improve the performance of UDFs, Apache Spark implemented support for using Apache Arrow to share data between processes. This means that the old-style pickling is no longer needed, and when we pass one column to our UDF, just that single column is passed between the processes. There are a few different names for UDFs that use Apache Arrow, and you may see them referred to as either Vectorized UDFs or Pandas UDFs. To create them in .NET for Apache Spark, instead of creating a function that takes native types, we need to use the columnar types such as Int64DataFrameColumn or StringDataFrameColumn which are defined in the Apache.Arrow NuGet package referenced by Microsoft.Spark.

Listings 4-9 and 4-10 show how to create a VectorUDF, which is very similar to a pickling UDF, except we need to define the function of type VectorUDF.

Listing 4-9. Using the DataFrameFunctions to create a VectorUDF in C#

```
using Microsoft.Data.Analysis;
using Microsoft.Spark.Sql;
using static Microsoft.Spark.Sql.DataFrameFunctions;

namespace Listing4_9
{
    class Program
    {
```

```
    static void Main(string[] args)
    {
        var spark = SparkSession.Builder().GetOrCreate();

        var dataFrame = spark.Sql("SELECT ID FROM range(1000)");

        var add100 = VectorUdf<Int64DataFrameColumn,
        Int64DataFrameColumn, Int64DataFrameColumn>((first, second) =>
        first.Add(second));

        dataFrame.Select(add100(dataFrame["ID"], dataFrame["ID"])).
        Show();
    }
  }
}
```

Listing 4-10. Using the DataFrameFunctions to create a VectorUDF in F#

```
open Microsoft.Data.Analysis
open Microsoft.Spark.Sql
open System

[<EntryPoint>]
let main argv =

    let spark = SparkSession.Builder().GetOrCreate();

    let dataFrame = spark.Sql("SELECT ID FROM range(1000)");

    let add100 = DataFrameFunctions.VectorUdf<Int64DataFrameColumn,
    Int64DataFrameColumn, Int64DataFrameColumn>(fun first second ->
    first.Add(second))

    dataFrame.Select(add100.Invoke(dataFrame.["ID"], dataFrame.["ID"])).
    Show()
    0
```

User-Defined Aggregate Functions (UDAFs)

UDAFs are similar to Vectorized UDFs in that they also use the Apache Arrow format, but instead of receiving an entire column and operating on that in a single operation, UDAFs work on grouped datasets, and each grouped set is sent to the UDAF to process. The result is that each group has a single output that we define.

If we look at Table 4-1, we have some data which we will group on the name column.

Table 4-1. *Sample data*

Name	Purchase	Cost
Ed	Sandwich	$4.95
Sarah	Drink	$2.95
Ed	Chips	$1.99
Ed	Drink	$3.45
Sarah	Sandwich	$8.95

We have two natural groups here, one for Ed and one for Sarah. If we create a UDAF group on the name column and pass in the Name and Cost columns, for example, then our UDAF would be called with an object called a RecordBatch, and there would be two RecordBatch's which would look like Tables 4-2 and 4-3.

Table 4-2. *First RecordBatch*

Name	Cost
Ed	$4.95
Ed	$1.99
Ed	$3.45

Table 4-3. *Second RecordBatch*

Name	Cost
Sarah	$2.95
Sarah	$8.95

What this allows us to do is to examine all of the columns that are passed in and create our aggregations. In Listing 4-11, we look at how we can process each of these batches using a UDAF in C#. After we have seen the example in C#, we will go through the example in F# in Listing 4-17.

Listing 4-11. How to process batches in a UDAF

```
static void Main(string[] args)
{
    var spark = SparkSession.Builder().GetOrCreate();

    var dataFrame = spark.Sql(
        "SELECT 'Ed' as Name, 'Sandwich' as Purchase, 4.95 as Cost UNION
        ALL SELECT 'Sarah', 'Drink', 2.95 UNION ALL SELECT 'Ed', 'Chips',
        1.99 UNION ALL SELECT 'Ed', 'Drink', 3.45  UNION ALL SELECT
        'Sarah', 'Sandwich', 8.95");

    dataFrame = dataFrame.WithColumn("Cost", dataFrame["Cost"].
    Cast("Float"));

    dataFrame.Show();
    var allowableExpenses = dataFrame.GroupBy("Name").Apply(new
    StructType(new[]
        {
            new StructField("Name", new StringType()),new StructField("Tota
            lCostOfAllowableExpenses", new FloatType())
        }), TotalCostOfAllowableExpenses
    );

    allowableExpenses.PrintSchema();
    allowableExpenses.Show();
}

private static RecordBatch TotalCostOfAllowableExpenses(RecordBatch records)
{
    var purchaseColumn = records.Column("Purchase") as StringArray;
    var costColumn = records.Column("Cost") as FloatArray;

    float totalCost = 0F;
```

```
    for (int i = 0; i < purchaseColumn.Length; i++)
    {
        var cost = costColumn.GetValue(i);
        var purchase = purchaseColumn.GetString(i);

        if(purchase != "Drink" && cost.HasValue)
            totalCost += cost.Value;
    }

    int returnLength = records.Length > 0 ? 1 : 0;

    return new RecordBatch(
        new Schema.Builder()
            .Field( f => f.Name("Name").DataType(ArrowStringType.Default))
            .Field( f => f.Name("TotalCostOfAllowableExpenses").
            DataType(Apache.Arrow.Types.FloatType.Default))
            .Build(),
        new IArrowArray[]
        {
            records.Column("Name"),
            new FloatArray.Builder().Append(totalCost).Build()
        }, returnLength);
}
```

If we break this down, in Listing 4-12, we create a DataFrame; this would typically be done by reading in some data. I then explicitly cast the Cost column to a float. When we work with the Apache Arrow format, every data type must be entirely correct. If there are any mistakes, the data will not be serialized and deserialized correctly, and Apache Spark will crash, so making sure that you know what data type each column is will help you.

Listing 4-12. Create a DataFrame and cast the cost column to float

```
var dataFrame = spark.Sql(
                "SELECT 'Ed' as Name, 'Sandwich' as Purchase, 4.95 as
                Cost UNION ALL SELECT 'Sarah', 'Drink', 2.95 UNION ALL
                SELECT 'Ed', 'Chips', 1.99 UNION ALL SELECT 'Ed', 'Drink',
                3.45  UNION ALL SELECT 'Sarah', 'Sandwich', 8.95");

dataFrame = dataFrame.WithColumn("Cost", dataFrame["Cost"].Cast("Float"));
```

In Listing 4-13, we see how we create a new DataFrame by calling the GroupBy function on the existing DataFrame. In our GroupBy call, we also define the structure of the DataFrame we will receive back from the UDAF. When I first started looking at UDAFs, I was under the impression that this was the schema of the DataFrame passed to the UDAF, but in fact it is the schema the UDAF will return.

Listing 4-13. Calling GroupBy on our existing DataFrame

```
var allowableExpenses = dataFrame.GroupBy("Name").Apply(new
StructType(new[]
            {
                new StructField("Name", new StringType()),new StructFie
                ld("TotalCostOfAllowableExpenses", new FloatType())
            }), TotalCostOfAllowableExpenses
        );
```

In Listing 4-14, we can see how we have received a RecordBatch in our UDAF, and we can retrieve any columns we need.

Listing 4-14. Retrieving rows from the RecordBatch

```
var purchaseColumn = records.Column("Purchase") as StringArray;
var costColumn = records.Column("Cost") as FloatArray;
```

What we have when we retrieve the columns are arrays, which we can iterate to example; remember that the rows we receive are for each unique group.

In Listing 4-15, we have our custom processing, and you can do whatever processing makes sense for your application at this point. In this example, we iterate through all of the rows and sum the cost of any purchase that is not a "Drink".

Listing 4-15. Processing RecordBatch's to include our custom logic

```
float totalCost = 0F;

for (int i = 0; i < purchaseColumn.Length; i++)
{
    var cost = costColumn.GetValue(i);
    var purchase = purchaseColumn.GetString(i);
```

```
    if(purchase != "Drink" && cost.HasValue)
        totalCost += cost.Value;
}
```

The critical thing to remember is that we don't have to care about the name column at all. The grouping is all handled by Apache Spark so we can keep a running total. This means that it is simple to calculate different values for each group, but conversely, it means that we are unable to share state between different groups.

In Listing 4-16, we return the data for the RecordBatch. In this case, we return the name and the total cost of allowable items. You can return anything you like here, but you will not be able to return more than one row per group.

Listing 4-16. Returning data to Apache Spark from the UDAF

```
int returnLength = records.Length > 0 ? 1 : 0;

            return new RecordBatch(
                new Schema.Builder()
                    .Field( f => f.Name("Name").DataType(ArrowStringType.
                    Default))
                    .Field( f => f.Name("TotalCostOfAllowableExpenses").
                    DataType(Apache.Arrow.Types.FloatType.Default))
                    .Build(),
                new IArrowArray[]
                {
                    records.Column("Name"),
                    new FloatArray.Builder().Append(totalCost).Build()
                }, returnLength);
        }
```

The data we return is in the Apache Arrow format, and as such, we need to use the Schema.Builder to create the schema and fields and then pass in the data as an array of IArrowArray objects. In this example, for the data, we pass in the name column precisely as it was passed to us, but for the cost column, we create a new FloatArray using the Builder and append the total cost. Following this pattern means that we receive many rows for each item in the group but return a single row, the aggregated data.

This can be quite confusing at first, but the key points are that for a UDAF, you receive a set of records for each group, each group at a time, and you return a single record for all the entries in the group you are processing.

In Listing 4-17, we have the same example but in F#.

Listing 4-17. User-Defined Aggregate Function in F#

```
open Apache.Arrow
open Apache.Arrow.Types
open Microsoft.Spark.Sql
open Microsoft.Spark.Sql.Types

let totalCostOfAllowableItems(records: RecordBatch): RecordBatch =

    let nameColumn  = records.Column "Name" :?> StringArray
    let purchaseColumn = records.Column "Purchase" :?> StringArray
    let costColumn = records.Column "Cost" :?> FloatArray

    let shouldInclude (purchase) = purchase <> "Drink"

    let count() =
        let mutable costs : float32 array  = Array.zeroCreate
        purchaseColumn.Length
        for index in 0 .. purchaseColumn.Length - 1 do
            costs.SetValue((if shouldInclude (purchaseColumn.
            GetString(index)) then costColumn.GetValue(index).Value else
            float32(0)), index)

        costs |> Array.sum

    let returnLength = if records.Length > 0 then 1 else 0

    let schema = Schema.Builder()
                    .Field(
                        Field("Name", StringType.Default, true))
                    .Field(
                        Field("TotalCostOfAllowableExpenses", FloatType.
                        Default, true)
                        )
                    .Build()
```

```
    let data: IArrowArray[] = [|
        nameColumn
        (FloatArray.Builder()).Append(count()).Build()
    |]

    new RecordBatch(schema, data, returnLength)

[<EntryPoint>]
let main argv =

    let spark = SparkSession.Builder().GetOrCreate();

    let dataFrame = spark.Sql("SELECT 'Ed' as Name, 'Sandwich' as Purchase,
    4.95 as Cost UNION ALL SELECT 'Sarah', 'Drink', 2.95 UNION ALL SELECT
    'Ed', 'Chips', 1.99 UNION ALL SELECT 'Ed', 'Drink', 3.45   UNION ALL
    SELECT 'Sarah', 'Sandwich', 8.95")
    let dataFrameWithCost = dataFrame.WithColumn("Cost", dataFrame.
    ["Cost"].Cast("Float"))

    dataFrameWithCost.Show()

    let structType = StructType ([|
        StructField("Name", StringType())
        StructField("TotalCostOfAllowablePurchases", FloatType())
        |])

    let categorized = dataFrameWithCost.GroupBy("Name").Apply(structType,
    totalCostOfAllowableItems)
    categorized.PrintSchema();
    categorized.Show();

    0
```

Debugging User-Defined Functions

Because we have this concept of Apache Spark starting a separate process or processes to handle the processing of the User-Defined Functions, it means that it is hard to debug in Visual Studio. The .NET for Apache Spark project includes the Microsoft.Spark.Worker process that, when triggered, calls the .NET Debugger.Launch() method, which pauses the process and displays the prompt to attach a debugger where you can choose your Visual Studio instance. Unfortunately, although I cannot find any documentation that confirms this, I am unable to get the `Debugger.Launch()` method to do anything on macOS or Linux, so unless you are on Windows, you may find that it isn't possible to debug UDFs or UDAFs in a debugger. Instead, you will need to fall back to doing things like creating log files and writing out details to disk. It isn't even any use using `Console.WriteLine()` as the output is swallowed up by Apache Spark and not displayed.

To enable the `Debugger.Launch()`, you can either add it to your UDF or UDAF code, which will trigger the UI to allow you to select your debugger, or you can set the environment variable "DOTNET_WORKER_DEBUG" to 1. When the worker process starts, if the environment variable exists and is set to 1, then the worker process calls `Debugger.Launch()` for you.

Summary

There is quite a lot to take in, just regarding User-Defined Functions and User-Defined Aggregate Functions. The key takeaways are that where possible, we should entirely avoid our user code for functions. If we can complete our processing without having to proxy any data at all, then that will be the quickest and the simplest. Where we care about performance but require our custom code, then we should avoid the old pickling UDFs and ensure we use the VectorUDF class.

CHAPTER 5

The DataFrame API

In this chapter, we will be having a look at the DataFrame API, which is the core API that we will use with .NET for Apache Spark. Apache Spark has a couple of different APIs, the Resilient Distributed Dataset (RDD) and DataFrame APIs, for processing. We will cover what the APIs are and why the RDD API is not available in .NET and that it is fine; the DataFrame API gives us everything we need.

The RDD API vs. the DataFrame API

The Resilient Distributed Dataset (RDD) API provides access to RDDs. RDDs are an abstraction over what could be massive data files by partitioning the files and spreading the processing over different compute nodes. When Apache Spark first came out, the RDD API was the only API available, and to use Apache Spark was to use the RDD API.

The DataFrame API is a higher-level abstraction and is based on columns of data, distributed on top of RDDs. The Column object includes lots of methods that we can use to write data processing code more efficiently. In Listing 5-1, we have a Scala RDD sample that parses an Apache web server log, groups the data by the user column, and sums the number of bytes and counts the number of requests. The full example comes with the Apache Spark installation in the file examples/src/main/scala/org/apache/spark/examples/LogQuery.scala. The sample uses map, reduceByKey, and collect to apply a UDF to the RDD. In Listing 5-2, we have a C# version of the example that, instead of using the RDD API, uses the DataFrame API and methods like GroupBy, Agg, Sum, and Count.

Listing 5-1. Example Scala program using the RDD API

```scala
object LogQuery {
  def main(args: Array[String]) {

    val sparkConf = new SparkConf().setAppName("Log Query")
    val sc = new SparkContext(sparkConf)
```

© Ed Elliott 2021
E. Elliott, *Introducing .NET for Apache Spark*, https://doi.org/10.1007/978-1-4842-6992-3_5

```scala
val dataSet =
  if (args.length == 1) sc.textFile(args(0)) else sc.parallelize
  (exampleApacheLogs)
// scalastyle:off
val apacheLogRegex =
  """^([\d.]+) (\S+) (\S+) \[([\w\d:/]+\s[+\-]\d{4})\] "(.+?)" (\d{3})
  ([\d\-]+) "([^"]+)" "([^"]+)".*""".r
// scalastyle:on
/** Tracks the total query count and number of aggregate bytes for a
particular group. */
class Stats(val count: Int, val numBytes: Int) extends Serializable {
  def merge(other: Stats): Stats = new Stats(count + other.count,
  numBytes + other.numBytes)
  override def toString: String = "bytes=%s\tn=%s".format(numBytes,
  count)
}

def extractKey(line: String): (String, String, String) = {
  apacheLogRegex.findFirstIn(line) match {
    case Some(apacheLogRegex(ip, _, user, dateTime, query, status,
    bytes, referer, ua)) =>
      if (user != "\"-\"") (ip, user, query)
      else (null, null, null)
    case _ => (null, null, null)
  }
}

def extractStats(line: String): Stats = {
  apacheLogRegex.findFirstIn(line) match {
    case Some(apacheLogRegex(ip, _, user, dateTime, query, status,
    bytes, referer, ua)) =>
      new Stats(1, bytes.toInt)
    case _ => new Stats(1, 0)
  }
}
```

```
    dataSet.map(line => (extractKey(line), extractStats(line)))
      .reduceByKey((a, b) => a.merge(b))
        .collect().foreach{
          case (user, query) => println("%s\t%s".format(user, query))}

    sc.stop()
  }
}
// scalastyle:on println
```

Listing 5-2. The same example rewritten for the DataFrame API in C#

```
static void Main(string[] args)
{
    Console.WriteLine("Hello World!");
    var regex = @"^([\d.]+) (\S+) (\S+) \[([\w\d:/]+\s[+\-]\d{4})\]
    ""(.+?)"" (\d{3}) ([\d\-]+) ""([^""]+)"" ""([^""]+)"".*";
    var spark = SparkSession.Builder().AppName("LogReader").GetOrCreate();
    var dataFrame = spark.Read().Text("log.txt");

    dataFrame
        .WithColumn("user", RegexpExtract(dataFrame["value"], regex, 3))
        .WithColumn("bytes", RegexpExtract(dataFrame["value"], regex, 7)
        .Cast("int"))
        .WithColumn("uri", RegexpExtract(dataFrame["value"], regex, 5))
        .Drop("value")
        .GroupBy("user", "uri")
        .Agg(Sum("bytes").Alias("TotalBytesPerUser"), Count("user").
        Alias("RequestsPerUser"))
        .Show();
}
```

The output for Listing 5-1:

```
(10.10.10.10,"FRED",GET http://images.com/2013/Generic.jpg
HTTP/1.1)  bytes=621  n=2
```

The output for Listing 5-2 shows the same data; however, the output is a DataFrame, which I call Show on to display the data rather than the data being native objects (strings, ints, etc.) in Scala.

```
+------+-------------------+----------------+--------------+
| user|                uri|TotalBytesPerUser|RequestsPerUser|
+------+-------------------+----------------+--------------+
|"FRED"|GET http://images...|             621|             2|
+------+-------------------+----------------+--------------+
```

There are two essential things to know about the RDD API when dealing with .NET for Apache Spark. The first is that the RDD API is not available from .NET, and there are no plans to make them available. It is possible that the RDD could be implemented in .NET, but to use the RDD API from .NET would mean having to write mostly using the pickling UDFs that we saw in the last chapter, and this would be slow. The second thing is that we cannot ignore the RDD API because the DataFrame API and the methods such as GroupBy and Agg are an abstraction over the RDD API. When you call the DataFrame API, the code makes RDD calls, and it is the RDD API that executes on the cluster.

In the Apache Spark 1.x timeframe, the performance difference between Python and Scala/Java was large because every operation that you wanted to do in Python required each row to be "pickled" or serialized/deserialized to Python for processing. In the Apache Spark 2.x timeframe, the DataFrame API meant that Python programs could call Scala methods on Column, which in turn called the RDD functions and left the data on the Java Virtual Machine (JVM) side. Leaving the data on the JVM side meant that the performance difference between Python and Scala/Java was very similar. It is this DataFrame API which makes it possible to write .NET code with similar performance, so it does make sense that the RDD API is not available in .NET for Apache Spark as the performance would not be great and the developer experience not as great.

Actions and Transformation

Before we examine the DataFrame API and dive deeper into all the possible things we can do with the DataFrame API, we must understand the difference between an action and a transformation. A transformation is something that will potentially be applied to a DataFrame, whereas an action applies all the previous transformations to a DataFrame. In Listing 5-3, I show a Spark SQL statement that will fail at runtime, but because there is no action, the program completes successfully.

Listing 5-3. A successfully completing query that should fail

```
spark.Sql("select assert_true(false)")
```

Passing false to `assert_true` should make the program crash, but when we run it, the program completes. If we add an action something like `Show`, `Collect`, `Take`, `Count`, `First`, then when we execute the program, we would get a failure. Listing 5-4 shows the same statement with an action that causes the runtime evaluation and subsequent failure.

Listing 5-4. An action terminates the statement, which causes the application to crash

```
spark.Sql("select assert_true(false)").Show()
```

When the action executes, there is a failure. However, the exception shows that the method that errored is `"showString"`:

```
Unhandled exception. System.Exception: JVM method execution failed:
Nonstatic method 'showString' failed for class '6' when called with
3 arguments ([Index=1, Type=Int32, Value=20], [Index=2, Type=Int32,
Value=20], [Index=3, Type=Boolean, Value=False], )
```

When you get an error, and the details of the exception show a method, it can often be a distraction from the original cause of the error, so it is important to realize that when you call a transformation, it may or may not be correct.

It may well feel at this point that it is impossible to debug a large failed Apache Spark application. However, it is not all bad. There are certain operations that are validated, such as the schema of a file or the columns in a query. Even though Listing 5-5 has no actions, there will still be a failure because the column that we are trying to use does not exist.

Listing 5-5. Failure will occur without an action under certain circumstances

```
spark.Sql("SELECT ID FROM Range(100)").Select("UnknownColumn")
```

In this case, enough of the Spark SQL statement is executed so that Apache Spark knows that the column UnknownColumn is not valid, so it will fail with an exception.

The DataFrame API

In this section, we will start to explore further what it is we can do with the DataFrame API. There are some fairly well-defined classes that should explore so that we can be effective with the DataFrame API. We will start with the DataFrameReader which is the class we use to read data into Apache Spark, and then we will look at how to create DataFrames without reading data and then the DataFrameWriter, which is how we write the results of our processing back out again, and finish with a more detailed look at the Column object, which is critical really when processing data in Apache Spark.

DataFrameReader

The DataFrameReader is the class that allows us to read from files and data sources, which we can then process with Apache Spark. We get to the DataFrameReader using a SparkSession, and Listing 5-6 shows how to read from the SparkSession and create a DataFrame, and in Listing 5-7, we show a DataFrameReader in F#.

Listing 5-6. Using the DataFrameReader to read data in C#

```
var spark = SparkSession.Builder().GetOrCreate();

DataFrameReader reader =
    spark.Read().Format("csv").Option("header", true).Option("sep", ",");

var dataFrame = reader.Load("./csv_file.csv");

dataFrame.Show();
```

Listing 5-7. Using the DataFrameReader to read data in F#

```
let spark = SparkSession.Builder().GetOrCreate()
let reader = spark.Read()
            |> fun reader -> reader.Format("csv")
            |> fun reader -> reader.Option("header", true)
            |> fun reader -> reader.Option("sep", "|")

let dataFrame = reader.Load("./csv_file.csv")
dataFrame.Show()
```

In Apache Spark, it is important to understand that many, if not all, of the objects we create such as a DataFrameReader are immutable, so if you did something like in Listing 5-8, we would not be modifying the original object and would not get the desired effect.

Listing 5-8. Each object is immutable, so unless we use method chaining, we could reference the wrong object

```
var spark = SparkSession.Builder().GetOrCreate();
var reader = spark.Reader();
reader.Option("header", true);
reader.Option("sep", "|");
reader.Csv("path.csv").Show();
```

If we ran this code, the options that we set on the reader are lost. If we wanted to keep them, then we should use method chaining such as in Listing 5-6.

CSV, Parquet, Orc vs. Load

There are two ways to ask Apache Spark to read a file in physically; the first is to use the Load method on the DataFrameReader, and the second is to call Format() and then Load(). Listing 5-9 shows you how to call the format-specific methods, and Listing 5-10 shows how to specify the format and call Load.

Listing 5-9. Using the custom format methods on the DataFrameReader

```
spark.Read().CSV("/path/to/CSV")
```

Listing 5-10. Specifying the format of the file and using Load

```
Spark.Read().Format("csv").Load("/path/to/.csv")
```

Each of these ways of using the DataFrameReader ends up with the same result, so you can choose which to use. I typically use the Format/Load approach when I do not know when writing the code which data format I will be loading or if the type information is going to be provided at runtime, possibly as some metadata that we receive with the file.

By default, the format is set to parquet, so if you want to use the Load method, then you either need to load a parquet file or call Format first. Natively Apache Spark supports the formats listed in Table 5-1. Still, it is possible to add JAR files for other data sources and load the file using the Format/Load method. An example is shown in Listing 5-11, which uses the Format method to specify an avro file, and in Listing 5-12, which uses the Format method to specify an Excel XLSX file.

Table 5-1. *Native file type support in Apache Spark*

File Type
Text
JSON
Parquet
ORC
JDBC

There are two things to note. Firstly, each of these file types has a method on the DataFrameReader such as Text(), JSON(), Parquet(), and so on, which allows files to be loaded. Secondly, although Java Database Connectivity (JDBC) is a way to connect to databases akin to ODBC or ADO.NET, we still consider it a file type in Apache Spark as the DataFrameReader object uses it, and there is no difference between using it and using a DataFrameReader to get data from a file.

Listing 5-11. Using Format to read from an avro file

```
spark.Format("avro").Load("/path/to/avro.avro");
```

Listing 5-12. Using Format to read from an Excel XLSX file

```
spark.Format("avro").Load("com.crealytics.spark.excel")
```

Because neither of these formats ship with Apache Spark, we would need to pass additional parameters to Apache Spark to tell it to load the JAR files that contain the code that can handle these two formats.

DataFrameReader Options

When reading files, there are a lot of considerations and options that we can use to load files. For example, in a CSV file, what is the file separator, and is there a header row or not? To see which options are available for which file type, you can visit the Apache Spark documentation for `DataFrameReader` and visit the file type methods such as `csv`, `text`, `json`, and `parquet`, and the method description contains a list of available options: `https://spark.apache.org/docs/latest/api/java/org/apache/spark/sql/DataFrameReader.html#csv-scala.collection.Seq-`.

The documentation shows the available options as well as showing the defaults. If we look at CSV for an example, we can see that today there are 28 specific options; the default separator (sep) is "," and the default encoding is "UTF-8".

When setting options, you can either specify them one by one or pass in a `"Dictionary<string, string>"` of all the options you wish to pass in. When you need to pass in a value type that isn't a string, use the string representation such as "true", and it will be converted to the correct type by Apache Spark.

Infer Schema vs. Manually Specified Schema

Some formats such as avro or parquet include the schema as a well-defined piece of metadata, alongside the data. Other file formats such as JSON or CSV do not include a schema definition, so Apache Spark, can attempt to infer the schema or not. For CSV files, the default is not to infer the schema, but you can enable inferring the schema using `Option("inferSchema", "true")`. JSON files will always attempt to infer the schema unless a schema is manually specified. It is not possible to not pass in a schema and also have Apache Spark not infer the schema.

There are two ways we can pass a schema to the DataFrameReader, and the first is to pass what we would call DDL in a traditional database system like SQL Server or Oracle; we demonstrate that in Listing 5-13. The second approach is to pass what is known as a `StructType`, which is the schema definition. If you do anything more complicated than a hello world type Apache Spark application, then you are likely to come across the `StructType` type again; Listing 5-14 shows how to pass a StructType to Apache Spark in C#, and Listing 5-15 shows how to pass a StructType to Apache Spark in F#.

Listing 5-13. Passing a DDL string to Apache Spark to specify the schema

```
var spark = SparkSession.Builder().GetOrCreate();
        var dataFrame = spark.Read().Option("sep", ",").
        Option("header", "false")
            .Schema("greeting string, first_number int, second_number
            float")
            .CSV("csv_file.csv");

        dataFrame.PrintSchema();
        dataFrame.
```

Running Listing 5-13 produces this output:

```
root
 |-- greeting: string (nullable = true)
 |-- first_number: integer (nullable = true)
 |-- second_number: float (nullable = true)

+--------+------------+-------------+
|greeting|first_number|second_number|
+--------+------------+-------------+
|   hello|         123|        987.0|
|      hi|         456|        654.0|
+--------+------------+-------------+
```

Listing 5-14. Passing a StructType to the DataFrameReader to manually specify the schema in C#

```
var spark = SparkSession.Builder().GetOrCreate();

var schema = new StructType(new List<StructField>()
{
    new StructField("greeting", new StringType()),
    new StructField("first_number", new IntegerType()),
    new StructField("second_number", new FloatType())
});
```

```
var dataFrame = spark.Read().Option("sep", ",").Option("header", "false")
    .Schema(schema)
    .Csv("csv_file.csv");

dataFrame.PrintSchema();
dataFrame.Show();
```

Listing 5-15. Passing a StructType to the DataFrameReader to manually specify the schema in F#

```
let dataFrame = SparkSession.Builder().GetOrCreate()
                |> fun spark -> spark.Read()
                |> fun reader ->
                    reader.Schema
                        (StructType
                            ([| StructField("greeting", StringType())
                                StructField("first_number", IntegerType())
                                StructField("second_number", FloatType())
                             |]))
                |> fun reader -> reader.Option("sep", ",").Option("header",
                    "false").Csv("csv_file.csv")

dataFrame.PrintSchema()
dataFrame.Show()
```

Creating DataFrames

There will be some rare cases where instead of reading data into Apache Spark, you want to create a DataFrame from code. There are a few approaches to creating DataFrames. We can call CreateDataFrame, and we can also use a SparkSession to run some Spark SQL that creates a data frame or use a SparkSession to create a DataFrame containing a sequential set of numbers using the Range method.

CreateDataFrame

The first approach is using CreateDataFrame, which we can pass it a list or an array of a specific type, which will create a single column made up of the values in the array or list. Alternatively, we can pass in an array or a list of GenericRow, which will allow us to create more than a single column. Listing 5-16 shows passing in arrays of single types, which creates a DataFrame with a single column, and also passing in a list of GenericRows, which also requires specifying the schema as a StructType, and Listing 5-17 shows how to create DataFrames using F#.

Listing 5-16. Creating DataFrames in C#

```
var spark = SparkSession.Builder().GetOrCreate();

spark.CreateDataFrame(new [] {"a", "b", "c"}).Show();
spark.CreateDataFrame(new [] {true, true, false}).Show();

var schema = new StructType(new List<StructField>()
{
    new StructField("greeting", new StringType()),
    new StructField("first_number", new IntegerType()),
    new StructField("second_number", new DoubleType())
});

IEnumerable<GenericRow> rows = new List<GenericRow>()
{
    new GenericRow(new object[] {"hello", 123, 543D}),
    new GenericRow(new object[] {"hi", 987, 456D})
};

spark.CreateDataFrame(rows, schema).Show();
```

Listing 5-17. Creating DataFrames using CreateDataFrame in F#

```
let spark = SparkSession.Builder().GetOrCreate()

spark.CreateDataFrame([| "a"; "b"; "c" |]).Show()

spark.CreateDataFrame([| true; true; false |]).Show()
```

```
spark.CreateDataFrame([| GenericRow([| "hello"; 123; 543.0 |])
                         GenericRow([| "hi"; 987; 456.0 |]) |],
                StructType
                    ([| StructField("greeting", StringType())
                        StructField("first_number", IntegerType())
                        StructField("second_number", DoubleType()) |]

                )).Show()
```

Running both Listings 5-16 and 5-17 emit the following output:

```
+---+
| _1|
+---+
|  a|
|  b|
|  c|
+---+

+-----+
|   _1|
+-----+
| true|
| true|
|false|
+-----+

+--------+------------+-------------+
|greeting|first_number|second_number|
+--------+------------+-------------+
|   hello|         123|        543.0|
|      hi|         987|        456.0|
+--------+------------+-------------+
```

Note that the first two DataFrames have a single column with the name "_1". In .NET for Apache Spark, the typed versions of CreateDataFrame convert the lists or arrays into GenericRows and create a StructType with the name "_1" and the data type of the list or array passed in before passing them over to Apache Spark.

Once you have your DataFrame, you can always rename the column using WithColumnRenamed and replace the existing column name with the new name. This is demonstrated in Listing 5-18.

Listing 5-18. Renaming a column using WithColumnRenamed

```
spark.CreateDataFrame(new [] {"a", "b", "c"}).WithColumnRenamed("_1",
"ColumnName").Show();
```

Spark SQL

The second approach to creating DataFrame is to pass some SQL to Apache Spark, and it will, if it can, create a DataFrame. Listing 5-19 shows some example SQL statements that could be passed to Apache Spark to generate a DataFrame.

Listing 5-19. Example Spark SQL statements that will generate DataFrames

```
var spark = SparkSession.Builder().GetOrCreate();
spark.Sql("SELECT ID FROM Range(100, 150)").Show();
spark.Sql("SELECT 'Hello' as Greeting, 123 as A_Number").Show();
spark.Sql("SELECT 'Hello' as Greeting, 123 as A_Number union SELECT 'Hi',
987").Show();
```

When this is run, the output shows all of the rows created:

```
+---+
| ID|
+---+
|100|
|101|
|102|
|103|
|104|
|105|
|106|
|107|
|108|
|109|
```

```
|110|
|111|
|112|
|113|
|114|
|115|
|116|
|117|
|118|
|119|
+---+
only showing top 20 rows

+--------+--------+
|Greeting|A_Number|
+--------+--------+
|   Hello|     123|
+--------+--------+

+--------+--------+
|Greeting|A_Number|
+--------+--------+
|   Hello|     123|
|      Hi|     987|
```

The Range Method

The last option for creating DataFrames from .NET is to use the Range method on the SparkSession object. The Range method takes a single integer value and Range returns a DataFrame with every value from 0 to the value passed in, or you can give in a start and end value, and you will get back a DataFrame with every value between the two values. Listing 5-20 shows the two usages of Range, and then we show the output from Listing 5-20.

Listing 5-20. Calling Range on the SparkSession to create a DataFrame

```
var spark = SparkSession.Builder().GetOrCreate();
spark.Range(5).Show();
spark.Range(10, 12).Show();
```

The output from this is

```
+---+
| id|
+---+
|  0|
|  1|
|  2|
|  3|
|  4|
+---+

+---+
| id|
+---+
| 10|
| 11|
+---+
```

DataFrameWriter

The DataFrameWriter is the class we use to write data back out again. It is similar to the DataFrameReader in that you can either write out using a specific format using Csv(), Parquet(), and so on or specify the format and use the Format() and Save() methods. We get to the DataFrameWriter from the DataFrame directly and show an example of writing a DataFrame in Listing 5-21.

Listing 5-21. The DataFrameWriter

```
var spark = SparkSession.Builder().GetOrCreate();
var dataFrame = spark.Range(100);

dataFrame.Write().Csv("output.csv");
dataFrame.Write().Format("json").Save("output.json");
```

The way that the DataFrameWriter works is very similar to the DataFrameReader. If you want to change the way the data is written, then there are a set of options that you can use. For example, when writing a CSV file, you can control the separator, the header, encoding, and so on. The best place to find all the available options you can set when writing a file is the DataFrameWriter documentation, and look at each of the methods for writing such as `csv()`, `json()`, and so on: `https://spark.apache.org/docs/latest/api/java/org/apache/spark/sql/DataFrameWriter.html#csv-java.lang.String-`.

DataFrameWriter Mode

When we write data, we can choose what happens if there is existing data. We can choose to append the data to add to the end of any existing data. We can choose to overwrite any existing data. We can choose to do nothing if data already exists, and finally we can choose to raise an error if data already exists. The last mode to error if data already exists is the default. In Listing 5-22, we show an example of all of the write modes.

Listing 5-22. Apache Spark DataFrameWriter write modes

```
var spark = SparkSession.Builder().GetOrCreate();
var dataFrame = spark.Range(100);

dataFrame.Write().Mode("overwrite").Csv("output.csv");
dataFrame.Write().Mode("ignore").Csv("output.csv");
dataFrame.Write().Mode("append").Csv("output.csv");
dataFrame.Write().Mode("error").Csv("output.csv");
```

Note that the last line will cause an exception because the file will already exist, and "error" throws an exception if the file already exists.

PartitionBy

When we write the data, we can also choose a column or columns to partition the data by. What this means is if we have a DataFrame that looks like Table 5-2 and we choose to partition by the year and country columns, we will end up with one file per year and for each country.

Table 5-2. *Sample Data*

Country	Year	Amount
UK	2020	500
UK	2020	1000
France	2020	500
France	1990	100
UK	1990	100

In Listing 5-23, the data is written, but partitioning on Country and Year, and what we end up with is five separate files written out, one file for every Country/Year combination. The path of the UK, 2020 file, for example, is "output.csv/Year=2020/Country=UK/part-randomguid.csv".

Listing 5-23. Partitioning the data when writing it out

```
var spark = SparkSession.Builder().GetOrCreate();
var dataFrame = spark.CreateDataFrame(new List<GenericRow>()
    {
        new GenericRow(new object[] {"UK", 2020, 500}),
        new GenericRow(new object[] {"UK", 2020, 1000}),
        new GenericRow(new object[] {"FRANCE", 2020, 500}),
        new GenericRow(new object[] {"FRANCE", 1990, 100}),
        new GenericRow(new object[] {"UK", 1990, 100})
    },
```

```
new StructType(
    new List<StructField>()
    {
        new StructField("Country", new StringType()),
        new StructField("Year", new IntegerType()),
        new StructField("Amount", new IntegerType())
    }
));
```

```
dataFrame.Write().PartitionBy("Year", "Country").Csv("output.csv");
```

If we partition data like this when we read and want to filter the data in Apache Spark, if we can filter on the partitioned columns, then reading is much more efficient. For example, if we used `"spark.Read().Csv("output.csv").Filter("Year = 2020 AND Country = 'UK'").Show();"`, then the partitions would be used so that only the data in the partitions matching the filter would be read in. If you have a lot of data but only need a small part of it, then this could make reading very efficient.

Controlling Filenames

When we write data using Apache Spark, and we specify a file and filename, such as "c:\temp\output.csv" or "/tmp/output.csv", what we will get is a folder called "output.csv" and inside that folder one or more files that follow the naming process of "part-partnumber-randomguid-jobid.format" such as "part-00003-de71ce5c-63aa-4bd9-863c-9696f9f86849.c000.csv".

The amount of individual files you end up with depends on how much data you have and how many partitions of that data you have. If you have to have just a single file, you can control how many files you end up with by doing a `Coalesce()` on the `DataFrame` *before* calling the `DataFrameWriter`. `Coalesce` will allow you to specify how many partitions you use when writing the data out.

It is not possible to control the filenames, and while this may be a bit confusing and annoying, it isn't a practical problem. We write data out, and when we read data back, we pass in the name of the folder, and Apache Spark will take care of finding any files in the directory or any subdirectories if partitioning is used.

Columns and Functions

The last part of the DataFrame API that we will cover in this chapter is the Column class. The Column class is what makes the DataFrame API so easy to use compared to the map/reduce type operation of the RDD API. Column is where the methods are available to actually process the data. Remember that the DataFrame API is based on columns of data, so it is natural that the Column class should be central to how we process data.

The Column class is a static member that belongs to Microsoft.Spark.Sql. Functions, and you can either get to Column using the Function class, such as Functions.Column, or use a static import in C# "using static Microsoft.Spark. Sql.Functions;". Column also has an alias Col, so if you see Column or Col, they are interchangeable.

Listings 5-24, C#, and 5-25, F#, show how we can use Column to process data in a DataFrame.

Listing 5-24. Using a Column or Col object in C#, including the static using statement to bring the Functions into scope

```
using Microsoft.Spark.Sql;
using static Microsoft.Spark.Sql.Functions;

namespace Listing5_24
{
    class Program
    {
        static void Main(string[] args)
        {
            var spark = SparkSession.Builder().GetOrCreate();
            var dataFrame = spark.Range(100);

            dataFrame.Select(Column("ID")).Show();
            dataFrame.Select(Col("ID")).Show();

            dataFrame.Select(Column("ID").Name("Not ID")).Show();
            dataFrame.Select(Col("ID").Name("Not ID")).Show();
```

```
            dataFrame.Filter(Column("ID").Gt(100)).Show();
            dataFrame.Filter(Col("ID").Gt(100)).Show();
        }
    }
}
```

Listing 5-25. Using a Column or Col object in F#

```
let spark = SparkSession.Builder().GetOrCreate()
let dataFrame = spark.Range(100L)

dataFrame.Select(Functions.Column("ID")).Show()
dataFrame.Select(Functions.Col("ID")).Show()

dataFrame.Select(Functions.Column("ID").Name("Not ID")).Show()
dataFrame.Select(Functions.Col("ID").Name("Not ID")).Show()

dataFrame.Filter(Functions.Column("ID").Gt(100)).Show()
dataFrame.Filter(Functions.Col("ID").Gt(100)).Show()
```

When we want to access a column, we can use `Function.Col` or `Function. Column`, or the DataFrame itself can be indexed using the column name such as `dataFrame["ColumnName"]`.

To see exactly what you can do with a Column and which Functions are available in .NET for Apache Spark, you can visit the documentation pages which are available for the column (`https://docs.microsoft.com/en-us/dotnet/api/microsoft.spark.sql. column?view=spark-dotnet`) and Functions (`https://docs.microsoft.com/en-us/ dotnet/api/microsoft.spark.sql.functions?view=spark-dotnet`).

Summary

The DataFrame API is core to how we read, process, and write data using Apache Spark. The DataFrame API is core to how we use Apache Spark from .NET, and understanding what a DataFrame is and how to read data in, process using Columns and Functions, and write the data back out again is core to how we programmatically work with Apache Spark.

In the next chapter, we will look at how we can access the power of Apache Spark using SQL queries through the use of hive tables. This different approach to accessing Apache Spark is part of its appeal to many. The people who want to program in Scala/Python/R/.NET can do that, and the people who want to use SQL can use that. I find myself writing mostly code but using SQL to explore the data or to help migrate existing legacy SQL solutions.

## CHAPTER 6

Spark SQL and Hive Tables

In this chapter, we are going to look at the Apache Spark SQL API. The SQL API allows us to write queries conforming to a subset of ANSI SQL:2003, which is the standard for the SQL database query language. The SQL API means we can store our data in files, probably in a data lake, and we can write SQL queries that access the data.

Before Apache Spark, Apache Hive was created by Facebook as a way to run SQL queries over data stored in Hadoop or even the Hadoop Distributed File System (HDFS). Apache Hive is made up of a "metastore" that is a set of metadata about files that allows developers to read them, as though they were tables in a database, and a query engine that converts SQL queries into map/reduce jobs that can be executed against the files stored in HDFS.

When Apache Spark was first released, it had the RDD API and no SQL support, but when Apache Spark 2.0 was released, it included a SQL parser and connectivity to the Apache Hive metastore. This meant that Apache Spark was able to run SQL queries using its own "catalyst" engine while using an Apache Hive metastore to store metadata needed to read from and to write to files.

What Is the SQL API

When we use the .NET for Apache Spark API, we generally have DataFrames. We either read them in from files or create new ones, but these are the unit of operation we work with, passing them to Apache Spark, transforming and writing back out again.

Instead of reading from or writing to files directly, we can use SQL to access any data we have stored where we have created metadata to point to those files.

© Ed Elliott 2021
E. Elliott, *Introducing .NET for Apache Spark*, https://doi.org/10.1007/978-1-4842-6992-3_6

In Listing 6-1, we will see how we can take a CSV file, register it as a table in the Hive metastore, and then use a SQL query to read the contents of that file in.

Listing 6-1. Create a table in the Hive metastore, pointing to a file on disk

```
var spark = SparkSession.Builder().GetOrCreate();
spark.Sql("CREATE TABLE Users USING csv OPTIONS (path './Names.csv')");

spark.Sql("SELECT * FROM Users").Show();
```

When we execute this program, we see the contents of the file:

```
» ./RunListing.sh 6 01
+------+
|  _c0|
+------+
|    Ed|
|  Bert|
|  Mary|
|Martha|
+------+
```

The Apache Spark DataFrame and SQL APIs have a similar feature to most modern database systems in that it will generate a plan, and there is a way to view the plan that is generated for how Apache Spark will execute the query. In Listing 6-2, we will look at the plan generated by running the preceding SQL statement and also by reading the same file using the DataFrame API, and we will see that the actual plans that are generated are the same.

Listing 6-2. Comparing plans from the SQL and DataFrame API

```
var spark = SparkSession.Builder().GetOrCreate();
spark.Sql("CREATE TABLE Users USING csv OPTIONS (path './Names.csv')");

spark.Sql("SELECT * FROM Users").Explain();
spark.Read().Format("csv").Load("./Names.csv").Explain();
```

When we execute this program, we see the following output:

```
== Physical Plan ==
*(1) FileScan csv default.users[_c0#10] Batched: false, Format: CSV,
Location: InMemoryFileIndex[file..., PartitionFilters: [], PushedFilters:
[], ReadSchema: struct<_c0:string>
== Physical Plan ==
*(1) FileScan csv [_c0#22] Batched: false, Format: CSV, Location:
InMemoryFileIndex[..., PartitionFilters: [], PushedFilters: [], ReadSchema:
struct<_c0:string>
```

Aside from the name of the table, which is included in the SQL version of the plan, they are the same, which shows whether you access your data through the DataFrame API or the SQL API. You still end up with the same execution path.

Passing Data Between Contexts

When we call SparkSession.Sql, the results are in a DataFrame, so to pass data from the SQL API over to your code where you can run your standard DataFrame calls is a matter of running a select statement. If we want to go the other way, that is, to take a DataFrame and make it available in the SQL context, then there is a step we need to make so that the data is recognized as a table in the hive catalog.

There are a number of ways to make a DataFrame available to SQL. The first is we can create what is called a managed table in Apache Hive. The DataFrameWriter object has a method called SaveAsTable. When we call this, the DataFrame is written as a set of parquet files and added to the Apache Hive catalog. Listings 6-3 and 6-4 show how to take a DataFrame and write it out as an Apache Hive managed table.

Listing 6-3. Writing a DataFrame as a managed Apache Hive table in C#

```
var spark = SparkSession.Builder().Config("spark.sql.legacy.
allowCreatingManagedTableUsingNonemptyLocation", "true").GetOrCreate();

var dataFrame = spark.CreateDataFrame(new [] {10, 11, 12, 13, 14, 15}).
WithColumnRenamed("_1", "ID");

dataFrame.Write().Mode("overwrite").SaveAsTable("saved_table");
spark.Sql("select * from saved_table").Show();
```

Listing 6-4. Writing a DataFrame as a managed Apache Hive table in F#

```
let spark = SparkSession.Builder().Config("spark.sql.legacy.
allowCreatingManagedTableUsingNonemptyLocation", "true").GetOrCreate()

    spark.CreateDataFrame([|10;11;12;13;14;15|])
        |> fun dataFrame -> dataFrame.WithColumnRenamed("_1", "ID")
        |> fun dataFrame -> dataFrame.Write().SaveAsTable("saved_table")

    spark.Sql("select * from saved_table").Show()
```

When we execute these programs, we can see the contents of the DataFrame:

```
+---+
| ID|
+---+
| 12|
| 15|
| 14|
| 13|
| 11|
| 10|
+---+
```

If we look at the output, we see that even though we created the DataFrame with the numbers in ascending order, they are now shown in a random order. This is because when we wrote the DataFrame out, it physically got persisted into a number of Parquet files, one file per partition. We get multiple files because of the way the work is split among the executioners, as we saw in Chapter 5.

If we look at the file system, the folder that I ran the program from has a spark-warehouse directory and inside that a folder with the same name as our table "saved_table" and finally a set of five Parquet files.

There is a second thing to notice, in that when I created the SparkSession, I had to pass in an option that would allow the SaveAsTable method to physically overwrite files in a directory if files already exist; just setting Mode("overwrite") on the DataFrameWriter is not enough. Note, though, that if you are using Apache Spark 3.0 or later, then you must remove this option as it will cause Apache Spark to throw an exception as the configuration setting is no longer valid.

I should point out here that the local spark-warehouse is from running the local instance of Apache Spark. In an environment, other than on your developer machine, we would have the Apache Hive warehouse configured properly, either with Databricks or AWS Glue includes an Apache Hive warehouse, or you can deploy and manage your own metastore. Of course, the end goal isn't to have a set of Parquet files on every developer's machine.

The next four ways to make a DataFrame available to a SQL query are the following methods on the DataFrame:

- **CreateTempView** – It creates a temporary view of the DataFrame. This fails if the view already exists. The temporary view is only available to the current SparkSession.

- **CreateOrReplaceTempView** – It creates a temporary view of the DataFrame. This does not fail if the view already exists. The temporary view is only available to the current SparkSession.

- **CreateGlobalTempView** – It creates a temporary view of the DataFrame. This fails if the view already exists. The temporary view is available to the current SparkSession and any other SparkSession on the cluster.

- **CreateOrReplaceGlobalTempView** – It creates a temporary view of the DataFrame. This does not fail if the view already exists. The temporary view is available to the current SparkSession and any other SparkSession on the cluster.

- Where the `DataFrameWriter.SaveAsTable` method was used to create a managed table in Apache Hive where the data was physically written as a set of parquet files. These create views over the existing data, so you do not need to go through the expense of writing the data to disk as an intermediate step.

- The variations of the different methods is to allow views to be read by other users of the Apache Spark instance, think in terms of a Databricks workspace where many users connect, and jobs run as different users, you can share data among the sessions. If the view is a global view, then when we select from it, we need to prefix it with the name of the global view database "global_temp", so if we create a global view called "global_temp_view", we can run this query to read from it in the SQL context: "select * from global_temp.global_temp_view".

- The difference between Create and CreateOrReplace controls whether you can overwrite an existing view or whether an exception is thrown if the view already exists.

In Listings 6-5 and 6-6, we show how to use these four functions in C# and F#.

Listing 6-5. Using the Create View methods on a DataFrame in C#

```
var spark = SparkSession.Builder().GetOrCreate();

var dataFrame = spark.CreateDataFrame(new [] {10, 11, 12, 13, 14, 15}).
WithColumnRenamed("_1", "ID");

dataFrame.CreateTempView("temp_view");
Console.WriteLine("select * from temp_view:");
spark.Sql("select * from temp_view").Show();

dataFrame.CreateOrReplaceTempView("temp_view");
Console.WriteLine("select * from temp_view:");
spark.Sql("select * from temp_view").Show();

dataFrame.CreateGlobalTempView("global_temp_view");
Console.WriteLine("select * from global_temp.global_temp_view:");
spark.Sql("select * from global_temp.global_temp_view").Show();

dataFrame.CreateOrReplaceGlobalTempView("global_temp_view");
Console.WriteLine("select * from global_temp.global_temp_view:");
spark.Sql("select * from global_temp.global_temp_view").Show();
```

Listing 6-6. Using the Create View methods on a DataFrame in F#

```
let spark = SparkSession.Builder().GetOrCreate()

let dataFrame = spark.CreateDataFrame([|10;11;12;13;14;15|]).
WithColumnRenamed("_1", "ID")

dataFrame.CreateTempView("temp_view")
printfn "select * from temp_view:"
spark.Sql("select * from temp_view").Show()

dataFrame.CreateOrReplaceTempView("temp_view")
printfn "select * from temp_view:"
spark.Sql("select * from temp_view").Show()

dataFrame.CreateGlobalTempView("global_temp_view")
printfn "select * from global_temp.global_temp_view:"
spark.Sql("select * from global_temp.global_temp_view").Show()

dataFrame.CreateOrReplaceGlobalTempView("global_temp_view")
printfn "select * from global_temp.global_temp_view:"
spark.Sql("select * from global_temp.global_temp_view").Show()

0
```

When running these two programs, they show the following output:

```
select * from temp_view:
+---+
| ID|
+---+
| 10|
| 11|
| 12|
| 13|
| 14|
| 15|
+---+
```

```
select * from temp_view:
+---+
| ID|
+---+
| 10|
| 11|
| 12|
| 13|
| 14|
| 15|
+---+

select * from global_temp.global_temp_view:
+---+
| ID|
+---+
| 10|
| 11|
| 12|
| 13|
| 14|
| 15|
+---+

select * from global_temp.global_temp_view:
+---+
| ID|
+---+
| 10|
| 11|
| 12|
| 13|
| 14|
| 15|
+---+
```

SparkSession Catalog

The SparkSession.Catalog is an object that allows us to inspect and modify the metadata that is stored in the Hive metastore. We can create tables and list the databases, tables, views, and functions and can also examine and drop those same objects. In Listings 6-7 and 6-8, we will create a new database using a SQL query and then query the database for a list of tables and use the catalog functions to examine the columns on a table. The listings use a parquet file that I have already generated; the parquet file includes three columns.

Listing 6-7. Working with Hive databases and tables in C#

```
var spark = SparkSession.Builder().GetOrCreate();
spark.Sql("CREATE DATABASE InputData");

spark.Catalog.SetCurrentDatabase("InputData");
spark.Catalog.CreateTable("id_list", "./ID.parquet");

var tables = spark.Catalog.ListTables("InputData");

foreach (var row in tables.Collect())
{
    var name = row[0].ToString();
    var database = row[1].ToString();

    Console.WriteLine($"Database: {database}, Table: {name}");
    var table = spark.Catalog.ListColumns(database, name);
    foreach (var column in table.Collect())
    {
        var columnName = column[0].ToString();
        var dataType = column[2].ToString();
        var nullable = (bool) column[3];
        var nullString = nullable ? "NULL" : "NOT NULL";

        Console.WriteLine($"{columnName}\t{dataType}\t{nullString}");
    }
}
```

Listing 6-8. Working with Hive databases and tables in F#

```fsharp
let spark = SparkSession.Builder().GetOrCreate()
spark.Sql("CREATE DATABASE InputData")
spark.Catalog.SetCurrentDatabase "InputData"
spark.Catalog.CreateTable("id_list", "./ID.parquet")

let getTableDefinition =
    let getColumn(column:Row) =
        sprintf "%s\t%s" (column.[0].ToString()) (column.[2].ToString())

    let getColumns(dbName:string, tableName:string) =
        spark.Catalog.ListColumns(dbName, tableName)
                        |> fun c -> c.Collect()
                        |> Seq.map(fun column -> getColumn(column))
                        |> String.concat "\n"

    let getTable (table:Row) =
        let databaseName = table.[1].ToString()
        let tableName = table.[0].ToString()

        let tableHeader = sprintf "Database: %s, Table: %s" databaseName
        tableName
        let columnDefinition = getColumns(databaseName, tableName)

        sprintf "%s\n%s" tableHeader columnDefinition

    let tableDefinition =
        spark.Catalog.ListTables "InputData"
        |> fun t -> t.Collect()
        |> Seq.map (fun table -> getTable(table))

    tableDefinition

PrettyPrint.print getTableDefinition

0
```

The output from these programs is

```
Database: inputdata, Table: id_list
Id        bigint
Age       bigint
halfAge   double
```

Being able to explore the objects is really useful in practice, and a real-life example of this was where I worked on a project which involved receiving data files from a number of source systems where the data schema we received could change without notice. What we developed was a way to compare the existing schema and the schema of the incoming file and determine if the schema could be evolved or if we would have to manually fix the schema to fix incompatibilities.

The ListDatabases, ListTables, and ListColumns methods are useful to explore which objects exist, but the catalog also has some other functions which check whether an object exists or not such as DatabaseExists, FunctionExists, and TableExists, which return a bool if the object exists.

The catalog also lets us drop temporary views we may have created using DropTempView and DropGlobalTempView.

If we know that an object exists, we can use GetDatabase, GetFunction, and GetTable which do not return DataFrames but instead first-class objects which give us access to their properties.

- **GetDatabase** – Returns a Database object that has the properties Description, Name, and LocationUri

- **GetFunction** – Returns a Function object that has the properties Database, Description, Name, ClassName, and IsTemporary

- **GetTable** – Returns a Table object that has the properties Database, Description, Name, IsTemporary, and TableType

Summary

In this chapter, we looked at how Apache Spark has an interface for running SQL queries, how we access DataFrames from those SQL queries, and how we can manage the metadata about the tables that are available to the SQL query engine.

Apache Spark SQL has a pretty feature-complete SQL parser and set of functions; to view the latest available functions, visit the Apache Spark documentation at `https://spark.apache.org/docs/latest/api/sql/`, and remember you can run the SQL queries from .NET using `SparkSession.Sql`.

## CHAPTER 7

Spark Machine Learning API

In this chapter, we will look at Spark's machine learning API or the MLLib API. The MLLib API is made up of both an RDD-based API and the newer DataFrame API. The DataFrame version of the API is referred to as the ML API because the objects exist in the org.apache.spark.ml namespace. From here on, we will use the term ML API to refer to the DataFrame version of the MLLib API. In the same way that the .NET for Apache Spark project supports the DataFrame API and not the RDD API, to date only the Spark ML API has any implementation.

The ML API was not part of the core project when it was first released and to date has only been implemented using external contributions, so it is not as complete as the rest of the APIs. Over time, the ML API will become more and more complete, but as of today, there are only a handful of ML objects that have been implemented. This means that we have a few different choices when writing machine learning applications with .NET for Apache Spark.

The first choice is that we use .NET purely and use the Microsoft library ML.NET, which means that you can create ML models using C# or F#. To access ML.NET from Apache Spark, we would then use a User-Defined Function (UDF) to pass the data over to the ML.NET models. This approach's drawback is that all the data has to be passed via UDFs, but this is probably the best choice today if you want to write all your code in .NET.

The second choice is that, if we do not have everything we need in .NET but can partially create or execute our model, we can do some of the work in .NET before saving our progress and calling into a Scala or Python Apache Spark program to read the output from .NET and finish the processing. This choice will be preferable where you already have existing models in Scala or Python, and you wish to migrate the code into .NET.

© Ed Elliott 2021
E. Elliott, *Introducing .NET for Apache Spark*, https://doi.org/10.1007/978-1-4842-6992-3_7

The last choice for implementing machine learning applications in .NET is to implement the objects you need yourself. Depending on what it is you are trying to implement, this can either be straightforward or can be quite hard to implement. In Appendix B, we show how to implement objects which you can use in your project or contribute back to the .NET for Apache Spark project.

Library Naming

To be specific about the naming, the original RDD-based machine learning API was named MLLib. With Apache Spark 2.0, the "Spark ML" library was created, which, although not an official name, was used to refer to the DataFrame API, and the objects in Scala were created in the org.apache.spark.ml package where the MLLib objects had previously been in the org.apache.spark.mllib. The MLLib API includes code for both the RDD API and the DataFrame API, but the objects we are implementing in .NET for Apache Spark will be, for now at least, from the org.apache.spark.ml package.

The thing to be careful of is when looking at the Apache Spark documentation; there will be objects with the same name in both packages, such as the org.apache.spark.mllib. feature.Word2Vec object which is separate to the org.apache.spark.ml.feature.Word2Vec object, which can lead to some confusion where you are expecting to see a set of parameters that do not exist on the MLLib RDD version of the object or vice versa.

Implemented Objects

The first set of objects to be created in the ML API are from the org.spark.ml.feature API and the implemented objects are

- Bucketizer

- CountVectorizer/CountVectorizerModel

- FeatureHasher

- HashingTF

- IDF/IDFModel

- Tokenizer

- Word2Vec/Word2VecModel

- SQLTransformer

- StopWordsRemover

There are several pending pull requests for more objects, and so I expect that this list will keep growing at, albeit a slow pace, a steady pace until we have feature parity between .NET for Apache Spark and Scala and Python.

To see the progress of the org.apache.spark.ml.feature objects which are being implemented, please see this issue at GitHub tracking the progress: `https://github.com/dotnet/spark/issues/381`.

Params

Parameters are a fundamental part of building machine learning applications. Understanding how to control a model to achieve the best possible outcome as well as knowing which parameters were used to build a model so that the model can be reproduced are essential for running machine learning applications in production. Without understanding how a decision was made using a machine learning application, there can be some severe consequences, including possible regulatory action. The EU GDPR law includes a specific section about machine learning called Article 22, which, among other things, consists of a note to say that it must be possible to provide the logic that was used to make a decision.

When we use the ML objects in Spark, each object typically takes a number of parameters, and there are two ways to access the parameters. The first is there is often a set of getters and setters on the object itself. For example, if we look at the `Word2Vec` object in Table 7-1, we can see the Get*, Set*, and parameter names we are working with for each parameter.

Table 7-1. *The getter/setter params on Word2Vec*

Get	Set	Parameter Name
GetInputCol	SetInputCol	inputCol
GetOutputCol	SetOutputCol	outputCol
GetVectorSize	SetVectorSize	vectorSize
GetMinCount	SetMinCount	minCount
GetMaxSentenceLength	SetMaxSentenceLength	maxSentenceLength
GetNumPartition	SetNumPartitions	numPartitions
GetSeed	SetSeed	Seed
GetStepSize	SetStepSize	stepSize
GetWindowSize	SetWindowSize	windowSize
GetVectorSize	SetVectorSize	vectorSize
GetMaxIter	SetMaxIter	maxIter

Practically, what this means is that we can use the provided getter and setter methods to control the parameters, or we can use the method Set and pass a parameter into object. In Listing 7-1, we show an example of how we can use the getter and setter methods or a `Param` object to set a specific parameter. We also introduce the `ExplainParams` method, which prints all of the available parameters, including any documentation, the current value, and what, if any, the default value is.

Listing 7-1. Controlling an object's parameters

```
var word2Vec = new Word2Vec();
word2Vec.SetSeed(123);

Console.WriteLine(word2Vec.ExplainParams());
```

Running this results in the following output:

```
inputCol: input column name (undefined)
maxIter: maximum number of iterations (>= 0) (default: 1)
```

maxSentenceLength: Maximum length (in words) of each sentence in the input data. Any sentence longer than this threshold will be divided into chunks up to the size (> 0) (default: 1000)
minCount: the minimum number of times a token must appear to be included in the word2vec model's vocabulary (>= 0) (default: 5)
numPartitions: number of partitions for sentences of words (> 0) (default: 1)
outputCol: output column name (default: w2v_cabb3eadcb81__output)
seed: random seed (default: -1961189076, current: 123)
stepSize: Step size to be used for each iteration of optimization (> 0) (default: 0.025)
vectorSize: the dimension of codes after transforming from words (> 0) (default: 100)
windowSize: the window size (context words from [-window, window]) (> 0) (default: 5)

We can see that the value of the seed has been set to 123. In Listing 7-2, we then use a Param object and the Set method to specify the parameter value.

Listing 7-2. Using a Param object to set a parameter value

```
var seedParam = new Param(word2Vec, "seed", "Setting the seed to 54321");
word2Vec.Set(seedParam, 54321L);

Console.WriteLine(word2Vec.ExplainParams());
```

Listing 7-2 results in the following output where we can see that the seed is now 54321:

inputCol: input column name (undefined)
maxIter: maximum number of iterations (>= 0) (default: 1)
maxSentenceLength: Maximum length (in words) of each sentence in the input data. Any sentence longer than this threshold will be divided into chunks up to the size (> 0) (default: 1000)
minCount: the minimum number of times a token must appear to be included in the word2vec model's vocabulary (>= 0) (default: 5)
numPartitions: number of partitions for sentences of words (> 0) (default: 1)
outputCol: output column name (default: w2v_cabb3eadcb81__output)
seed: random seed (default: -1961189076, current: 54321)

stepSize: Step size to be used for each iteration of optimization (> 0)
(default: 0.025)
vectorSize: the dimension of codes after transforming from words (> 0)
(default: 100)
windowSize: the window size (context words from [-window, window]) (> 0)
(default: 5)

Finally, in Listing 7-3, instead of creating a new Param object, we ask the Word2Vec object to give us a parameter named "seed" which we can then use to set the parameter.

Listing 7-3. Using a Param object supplied by the Word2Vec object to set a parameter value

```
var seed = word2Vec.GetParam("seed");
word2Vec.Set(seed, 12345L);
Console.WriteLine(word2Vec.ExplainParams());
```

We can see in the output that the parameter value has been set to 12345:

```
inputCol: input column name (undefined)
maxIter: maximum number of iterations (>= 0) (default: 1)
maxSentenceLength: Maximum length (in words) of each sentence in the input
data. Any sentence longer than this threshold will be divided into chunks
up to the size (> 0) (default: 1000)
minCount: the minimum number of times a token must appear to be included in
the word2vec model's vocabulary (>= 0) (default: 5)
numPartitions: number of partitions for sentences of words (> 0) (default: 1)
outputCol: output column name (default: w2v_cabb3eadcb81__output)
seed: random seed (default: -1961189076, current: 12345)
stepSize: Step size to be used for each iteration of optimization (> 0)
(default: 0.025)
vectorSize: the dimension of codes after transforming from words (> 0)
(default: 100)
windowSize: the window size (context words from [-window, window]) (> 0)
(default: 5)
```

One crucial point is that when we use the `Param` object and the `Set` method, the data types are not validated, so it is possible to set a parameter to an incorrect type. You will not know until you either save your object or try to use it. It is often safer than to use the supplied getters and setters on each object. The Scala version of the `Param` object has a way to verify parameters, but in .NET, we will just need to be careful.

If you ever want to reset a parameter back to its original default value, you can use the `Clear` method, as shown in Listing 7-4.

Listing 7-4. Clearing any parameters which have previously been set

```
var seed = word2Vec.GetParam("seed");
word2Vec.Set(seed, 12345L);
Console.WriteLine(word2Vec.ExplainParams());

word2Vec.Clear(seed);
Console.WriteLine(word2Vec.ExplainParams());
```

Saving/Loading Objects

Each of the core objects in the Spark.ML namespace includes a method called `Save` and a static method called `Load`. The `Load` and `Save` methods persist a copy of the object, including any runtime information, which then allow them to be read back into memory. This is especially useful with machine learning applications as we might want to create and train a model on one set of data, then save the objects so that they can be reused later to use the model or run a prediction. In Listing 7-5, we see the `Load` and `Save` methods being used. Note that although they are in the same process, the objects could be saved in one process and loaded into another. The language is irrelevant, so you could create a model in .NET, save it, then load and use it from Scala.

Listing 7-5. The Load and Save methods

```
bucketizer.SetInputCol("input_column");
bucketizer.Save("/tmp/bucketizer");

bucketizer.SetInputCol("something_else");
```

```
var loaded = Bucketizer.Load("/tmp/bucketizer");
Console.WriteLine(bucketizer.GetInputCol());
Console.WriteLine(loaded.GetInputCol());
```

When we run this, then we can see that the original Bucketizer, which had its inputColumn set to "something else," is still valid, but the new loaded-in Bucketizer has the original value of "input_column".

```
something_else
input_column
```

Identifiable

The objects also typically implement Identifiable, which means when you create a new one, you can optionally specify a unique string to identify the specific instance of the object. If you do not specify a unique string, then one is generated for you. You can use this unique string later on to identify the exact instance of the object. When you create Param objects, you need to identify the object to which the param will belong, and that is done either by passing in the object itself or passing the string identifier. In Listing 7-6, we show how to pass in a unique string to one of the Spark.ML objects and how to reference that unique string later on.

Listing 7-6. The uid of a Spark.ML object instance

```
var tokenizer = new Tokenizer();
Console.WriteLine(tokenizer.Uid());

tokenizer = new Tokenizer("a unique identifier");
Console.WriteLine(tokenizer.Uid());
```

The output of this is

```
tok_34a2ad14b80a
a unique identifier
```

TF-IDF

The implemented ML objects in .NET for Apache Spark are nowhere near parity to Spark's number of objects in Spark.ML; however, there is already enough functionality to run useful machine learning applications. In this section, we will build a working example of "term frequency, inverse document frequency" or TF-IDF, which is a way to search for some text in a set of documents and find relevant documents. TF-IDF is based on the fact that if you just do a wildcard search for terms, then you will find documents where the term exists but isn't very relevant. TF-IDF weighs how common a word is in one specific document compared to how many terms there are in all of the documents and how relevant those search terms are. For example, a book might have "Page xx" written on every page, but the term is not very relevant to the document. However, if there were a document talking about how pages were laid out, then the word page in that document would be very relevant.

TF-IDF is discussed in detail in this Wikipedia article: *https://en.wikipedia.org/ wiki/Tf%E2%80%93idf*. The high-level process to use TF-IDF is

1. *Get some documents to use as the source.*

2. *Read the documents into a DataFrame.*

3. *Use a Tokenizer to split the documents into words.*

4. *Use a HashingTF to build a vector containing a hash of each of the individual words.*

5. *Create an IDF and create an IDFModel by "Fitting" the hash of each of the words, that is, giving each word or term a frequency count and relative importance.*

6. *Get some search terms and turn them into a DataFrame.*

7. *Use the Tokenizer to split the search term into words.*

8. *Use the HashingTF to build a vector containing a hash of each of the individual words in the search term.*

9. *Use the IDFModel to transform the search terms to give them the same relative weights that were in the documents.*

10. *Join the datasets together and work out how close the search term*
is to the document by calculating the cosine similarity of both and
ordering the results by how closely they match. To understand why
we use cosine similarity with TF-IDF, have a look at this excellent
blog post: https://janav.wordpress.com/2013/10/27/tf-idf-
and-cosine-similarity/.

For this example, I will use *The Complete Works of William Shakespeare* and then
find relevant documents to specific search terms. There are a few sources of the data we
need, but I downloaded this repo (https://github.com/severdia/PlayShakespeare.
com-XML), including a copy of all the works in XML format, which made it simple to read
in the text and title of each poem or play as well as a permissive license.

The full example application is in Listing7-ExampleCSharp and Listing7-
ExampleFSharp.

In Listings 7-7 (C#) and 7-8 (F#), we read each of the documents in as XML and parse
the XML to retrieve the work's text and title.

Listing 7-7. Reading the contents of each work as XML and retrieving the
document and title in C#

```
private static List<GenericRow> GetDocuments(string path)
{
    var documents = new List<GenericRow>();

    foreach (var file in new DirectoryInfo(path).EnumerateFiles("*.xml",
    SearchOption.AllDirectories))
    {
        var doc = new XmlDocument();

        doc.Load(file.FullName);

        var playTitle = "";
        var title = doc.SelectSingleNode("//title");

        playTitle = title != null ? title.InnerText : doc.SelectSingleNode
        ("//personae[@playtitle]").Attributes["playtitle"].Value;

        var play = doc.SelectSingleNode("//play");
```

```
        if (play != null)
        {
            documents.Add(new GenericRow(new[] {playTitle, play.
            InnerText}));
        }
        else
        {
            var poem = doc.SelectSingleNode("//poem");
            documents.Add(new GenericRow(new[] {playTitle, poem.
            InnerText}));
        }
    }
}

    return documents;
}

var spark = SparkSession
    .Builder()
    .GetOrCreate();

var documentPath = args[0];
var search = args[1];

var documentData = GetDocuments(documentPath);
```

Listing 7-8. Reading the contents of each work as XML and retrieving the document and title in F#

```
let createXmlDoc(path: string) =
    let doc = XmlDocument()
    doc.Load(path)
    doc

let parseXml(doc: XmlDocument) =
    let selectSingleNode node =
        Option.ofObj (doc.SelectSingleNode(node))
```

```
let documentTitle =
    match selectSingleNode "//title" with
    | Some node -> node.InnerText
    | None -> doc.SelectSingleNode("//personae[@playtitle]").
      Attributes.["playtitle"].Value

match selectSingleNode "//play" with
| Some node -> GenericRow([|documentTitle; node.InnerText|])
| None -> GenericRow([|documentTitle; doc.SelectSingleNode("//poem").
  InnerText|])

let getDocuments path = System.IO.Directory.GetFiles(path, "*.xml")
                                |> Seq.map (fun doc -> createXmlDoc doc)
                                |> Seq.map (fun xml -> parseXml xml)

let main argv =
    let args = match argv with
                | [|documentPath; searchTerm|] -> {documentsPath = argv.
                  [0]; searchTerm = argv.[1]; success = true}
                | _ -> {success = false; documentsPath = ""; searchTerm = ""}

    match args.success with
        | false ->
            printfn "Error, incorrect args. Expecting 'Path to documents'
            'search term', got: %A" argv
            -1

        | true ->
            let spark = SparkSession.Builder().GetOrCreate()
            let documents = getDocuments args.documentsPath
```

Now that we have the documents read into our .NET application, we need to create a DataFrame so Apache Spark can work with the documents. An alternative to reading the files through .NET and then creating a DataFrame would be to get Apache Spark to read the XML files and process them or preprocess them with .NET and write them to disk in a friendlier format for Apache Spark such as Parquet or Avro. In this case, because there are about 50 documents, I will create a DataFrame and add the documents to it rather than writing the documents back out again. If there were thousands or millions of documents, then we would need to think of different approaches.

In Listings 7-9 and 7-10, we create a DataFrame, which involves passing an IEnumerable<GenericRow> and a schema describing our rows.

Listing 7-9. CreateDataFrame passing in our specific schema in C#

```
var documents = spark.CreateDataFrame(documentData, new StructType(new
List<StructField>
{
    new StructField("title", new StringType()),
    new StructField("content", new StringType())
}));
```

Listing 7-10. CreateDataFrame passing in our specific schema in F#

```
let documents = spark.CreateDataFrame(documents, StructType([|StructField("
title", StringType());StructField("content", StringType())|]))
```

The next thing we will do is to create the Spark.ML objects that we need. Table 7-2 lists the objects and what we will use them for.

Table 7-2. *Spark.ML objects we need for our machine learning application*

Object	Reason	Training	Execution
Tokenizer	Splits the documents into an array of words within the DataFrame	Yes	Yes
HashingTF	Converts words into numeric representations of each word	Yes	Yes
IDF	Uses the documents to build a model which describes how frequent terms are used across all of the documents	Yes	No Once the model has been "trained" with the sample dataset, we use that rather than retraining the model each time
IDFModel	This is the model that has been "Fit" with the documents and includes how frequent each term is across the entire set of documents	Yes	Yes

In Listings 7-11 and 7-12, we create the objects used in the initial training phase and the execution phase, the Tokenizer, HashingTF, and IDF. We will create the IDFModel using the actual documents.

Listing 7-11. Creating the Tokenizer, HashingTF, and IDF in C#

```
var tokenizer = new Tokenizer()
    .SetInputCol("content")
    .SetOutputCol("words");

var hashingTF = new HashingTF()
    .SetInputCol("words")
    .SetOutputCol("rawFeatures")
    .SetNumFeatures(1000000);

var idf = new IDF()
    .SetInputCol("rawFeatures")
    .SetOutputCol("features");
```

Listing 7-12. Creating the Tokenizer, HashingTF, and IDF in F#

```
let tokenizer = Tokenizer().SetInputCol("content").SetOutputCol("words")
let hashingTF = HashingTF().SetInputCol("words").
SetOutputCol("rawFeatures").SetNumFeatures(1000000)
let idf = IDF().SetInputCol("rawFeatures").SetOutputCol("features")
```

Each of the objects uses a DataFrame to process, so we need to tell the objects which columns to use. For example, to use the Tokenizer, we tell it that it will find its input data in the column "content," and it should write its output data into the "words" column, which the Tokenizer will create. The HashingTF will look in the "words" column for its input data and out the data into the "rawFeatures" column.

In Listings 7-13 and 7-14, we then take the documents and split them into individual words and then into Vectors, which are numeric identifiers for each word. We use numbers instead of strings because we need to run some calculations, specifically to calculate the cosine similarity of each document compared to our search term, and we cannot do that with strings.

Listing 7-13. Transforming the documents into words and vectors in C#

```
var tokenizedDocuments = tokenizer.Transform(documents);
var featurizedDocuments = hashingTF.Transform(tokenizedDocuments);
```

Listing 7-14. Transforming the documents into words and vectors in F#

```
let featurized = tokenizer.Transform documents
                            |> hashingTF.Transform
```

If we were to call the Show method on the DataFrame that the HashingTF returned, then it would look like

```
+---------+--------+---------+------------------+
|    title| content|   words|        rawFeatures|
+---------+--------+---------+------------------+
|The So...|The S...|[the, ...|(1000000,[522, ...|
|The Tw...|The T...|[the, ...|(1000000,[130, ...|
```

The content has been split into an array of words, and each word has been given a numeric identifier.

Now we have the documents in a format that we can work with. We need to "train" the model by "fitting" the documents to IDF. We show this in Listings 7-15 and 7-16.

Listing 7-15. "Fitting" the dataset to the IDF to create the model in C#

```
var idfModel = idf.Fit(featurizedDocuments);
```

Listing 7-16. "Fitting" the dataset to the IDF to create the model in F#

```
let model = featurized
              |> idf.Fit
```

Now we have the objects we need, and we have the model which has been trained over the dataset of documents we need to calculate, for each document, how large it is compared to all of the other documents. To do this, for each row in our DataFrame, that is, for each document, we iterate through every value in our dataset and square the number and then take the square root of the squares. In Listings 7-17 and 7-18, we iterate through every value in our Vectors and calculate the normalization value we will use later on when working out how similar each document is to our search term.

Listing 7-17. Calculating the normalization number to use later on in C#

```
private static readonly Func<Column, Column> udfCalcNorm = Udf<Row,
double>(row =>
    {
        var values = (ArrayList) row.Values[3];
        var norm = 0.0;

        foreach (var value in values)
        {
            var d = (double) value;
            norm += d * d;
        }

        return Math.Sqrt(norm);
    }
);

var transformedDocuments = idfModel.Transform(featurizedDocuments).
Select("title", "features");
            var normalizedDocuments = transformedDocuments.
            Select(Col("features"), udfCalcNorm(transformedDocuments
            ["features"]).Alias("norm"), Col("title"));
```

Listing 7-18. Calculating the normalization number to use later on in F#

```
let calcNormUDF = Functions.Udf<Row, double>(fun row -> row.Values.[3] :?>
ArrayList
                                        |> Seq.cast
                                        |> Seq.map (fun item
                                            -> item * item)
                                        |> Seq.sum
                                        |> Math.Sqrt)

let normalizedDocuments = model.Transform featurized
                                |> fun data -> data.Select
                                (Functions.Col("features"),
                                calcNormUDF.Invoke
                                (Functions.Col("features")).
                                Alias("norm"), Functions.
                                Col("title"))
```

Until version 1.0 of .NET for Apache Spark, it would not have been possible to bring the Vector from the JVM over to .NET, and even so, with version 1.0, the data supplied to the UDF is the internal representation of the Vector. In Apache Spark, there are two types of Vector, a DenseVector and, what we have here, a SparseVector. If you use one of the Vector types in Scala or Python, then you can use them and work with them as Vectors. Hopefully, at some point in the future, you will be able to work with Vectors in .NET for Apache Spark, but until then, we will need to be aware of how the Vectors are implemented internally.

The DenseVector is the easiest to work with as a double array backs it, that is, an array of doubles. The SparseVector which we have here is harder to work with because instead of having an array of all the values, any value that is 0.0 is excluded from the SparseVector, so if you wanted to represent 1.0, 2.0, 0.0, 4.0 as a SparseVector, what you would have is an array with a list of the index of each element; if the value is 0.0, then the index is omitted. In our SparseVector example, we would have two arrays, one with the following indexes 0, 1, 3 and a second array with the values 1.0, 2.0, 3.0. When we want to iterate the SparseVector, then we need to iterate each of the indexes. Where an index value is missing, we know the value is 0.0, but where the value is in the index, we use the position of the index to look up the actual value. In our example, if we wanted to see what the value at the fourth position in the SparseVector is, we would go to the index and search for the value 3; remember this is a zero-based array. The value 3 is at the third position in the array or index 2, which points to 3.0 in the values array.

In Table 7-3, we can see how the words have been split into tokens.

Table 7-3. *An example SparseVector and how to retrieve specific indexes.*

Vector	Index	Values	Value at Index 5
0.0, 0.0, 0.1, 0.0, 0.0, *0.2*	2, *5*	0.1, *0.2*	*0.2*
0.1, 0.2, 0.3, 0.4, 0.0, *0.0*	0, 1, 2, 3	0.1, 0.2, 0.3, 0.4	*0.0*

Practically, what this means is our UDF receives an array that consists of four objects, listed in Table 7-4.

Table 7-4. *The details of the SparseVector which is given to a UDF as an array of objects*

Index	Type	Description
0	Int	The starting offset for this application will always be 0
1	Int	How many items in the Vector this SparseVector represents. The SparseVector may contain ten values, but the SparseVector could represent millions of items in the Vector
2	Int	The indexes in the Vector which point to values that are not 0.0
3	Double	The values in the Vector which are not 0.0

In Listings 7-19 and 7-20, we take the search term, create a DataFrame, and then run the same process of splitting into words, creating a Vector and using the model to transform the vectors into a set of features we can compare with the original documents. The only thing we don't need to do is to rebuild the model because we have the model of the original documents, which we will have to reuse; otherwise, our search terms would have differing weights to the original documents.

Listing 7-19. Converting the search term into a DataFrame that can be compared with the original documents in C#

```
var searchTerm = spark.CreateDataFrame(
    new List<GenericRow> {new GenericRow(new[] {search})},
    new StructType(new[] {new StructField("content", new StringType())}));

var tokenizedSearchTerm = tokenizer.Transform(searchTerm);

var featurizedSearchTerm = hashingTF.Transform(tokenizedSearchTerm);

var normalizedSearchTerm = idfModel
    .Transform(featurizedSearchTerm)
    .WithColumnRenamed("features", "searchTermFeatures")
    .WithColumn("searchTermNorm",
udfCalcNorm(Column("searchTermFeatures")));
```

Listing 7-20. Converting the search term into a DataFrame that can be compared with the original documents in F#

```
let term = GenericRow([|"Montague and capulets"|])
let searchTerm = spark.CreateDataFrame([|term|], StructType([|StructField("
content", StringType())|]) )

tokenizer.Transform searchTerm
    |> hashingTF.Transform
    |> model.Transform
    |> fun data -> data.WithColumnRenamed("features", "searchTermFeatures")
    |> fun data -> data.WithColumn("searchTermNorm", calcNormUDF.
        Invoke(Functions.Col("searchTermFeatures")))
```

The final thing to do is to join the original documents and the search terms into a single DataFrame and calculate the cosine similarity of the two Vectors, which we show in Listings 7-21 and 7-22. We calculate the cosine similarity by multiplying every value in the Vector by the value in the second Vector that is in the same position in the array. We then divide the results by the product of the normalization of both the document and search term that we previously calculated. Note also that this is a SparseVector, so we need to do some work to identify the values at specific offsets.

Listing 7-21. Calculating the cosine similarity using C#

```
private static readonly Func<Column, Column, Column, Column, Column>
udfCosineSimilarity =
    Udf<Row, Row, double, double, double>(
        (vectorA, vectorB, normA, normB) =>
        {
            var indicesA = (ArrayList) vectorA.Values[2];
            var valuesA = (ArrayList) vectorA.Values[3];

            var indicesB = (ArrayList) vectorB.Values[2];
            var valuesB = (ArrayList) vectorB.Values[3];

            var dotProduct = 0.0;
```

```
        for (var i = 0; i < indicesA.Count; i++)
        {
            var valA = (double) valuesA[i];

            var indexB = findIndex(indicesB, 0, (int) indicesA[i]);

            double valB = 0;
            if (indexB != -1)
            {
                valB = (double) valuesB[indexB];
            }
            else
            {
                valB = 0;
            }

            dotProduct += valA * valB;
        }

    var divisor = normA * normB;

    return divisor == 0 ? 0 : dotProduct / divisor;
});
```

Listing 7-22. Calculating the cosine similarity using F#

```
let cosineSimilarity (vectorA:Row, vectorB:Row, normA:double,
normB:double):double =

    let indicesA = vectorA.Values.[2]  :?> ArrayList
    let valuesA = vectorA.Values.[3] :?> ArrayList

    let indicesB = vectorB.Values.[2] :?> ArrayList
    let valuesB = vectorB.Values.[3] :?> ArrayList

    let indexedA = indicesA |> Seq.cast |> Seq.indexed
    let indexedB = indicesB |> Seq.cast |> Seq.indexed |> Seq.map
    (fun item -> (snd item, fst item)) |> Map.ofSeq

    PrettyPrint.print indexedB
```

```
let findIndex value = match indexedB.ContainsKey value with
                        | true -> indexedB.[value]
                        | false -> -1

let findValue indexA =
                    let index =  findIndex indexA

                    match index with
                        | -1 -> 0.0
                        | _ -> unbox<double> (valuesB.
                          Item(unbox<int> (index)))

let dotProduct = indexedA
                    |> Seq.map (fun index -> (unbox<double>valuesA.[fst
                       index]) * (findValue (unbox<int> indicesA.[fst
                       index])))
                    |> Seq.sum

normA * normB |> fun divisor -> match divisor with
                                        | 0.0 -> 0.0
                                        | _ -> dotProduct / divisor

let cosineSimilarityUDF = Functions.Udf<Row, Row, double, double,
double>(fun vectorA vectorB normA normB -> cosineSimilarity(vectorA,
vectorB, normA, normB))
```

In Listings 7-23 and 7-24, we have the final step of joining the DataFrames, calculating the cosine similarity, ordering the results by most similar to least similar, and then printing out the title and the similarity.

Listing 7-23. Joining the DataFrames and calculating the cosine similarity to generate our best matching results in C#

```
var results = normalizedDocuments.CrossJoin(normalizedSearchTerm);

results
    .WithColumn("similarity", udfCosineSimilarity(Column("features"),
    Column("searchTermFeatures"), Col("norm"), Col("searchTermNorm")))
    .OrderBy(Desc("similarity")).Select("title", "similarity")
    .Show(10000, 100);
```

Listing 7-24. Joining the DataFrames and calculating the cosine similarity to generate our best matching results in F#

```
|> normalizedDocuments.CrossJoin
|> fun data -> data.WithColumn("similarity", cosineSimilarityUDF.
Invoke(Functions.Col("features"), Functions.Col("searchTermFeatures"),
Functions.Col("norm"), Functions.Col("searchTermNorm")))
|> fun matched -> matched.OrderBy(Functions.Desc("similarity")).
Select("title", "similarity")
|> fun ordered -> ordered.Show(100, 1000)
```

In Table 7-5, I ran the program with various search terms, and these were the results, which I think are surprisingly accurate.

Table 7-5. *Search terms and their results when run against the complete works of Shakespeare*

Search Term	Position	Title	Similarity
"lovers in the woods poison themselves"	1	A Midsummer Night's Dream	0.04105529867838565
	2	As You Like It	0.02845396350570514
	3	Love's Labour's Lost	0.0141767769638970023
"witches dagger blood on hands."	1	The Tragedy of Macbeth	0.08824800070165366
	2	The Comedy of Errors	0.0259932970399907045
	3	The Second Part of Henry the Sixth	0.007198784643312808

Summary

In this chapter, we have looked at the Spark.ML API, which, although it is not anywhere near as feature-complete as the rest of the .NET version of the Apache Spark APIs, is still useful and is actively being developed to increase the coverage.

There are some intricacies to using the Spark.ML API in .NET, such as having to work with the raw SparseVector, but hopefully these barriers to entry should be removed soon.

PART III

Examples

## CHAPTER 8

Batch Mode Processing

In this chapter, we will be looking at how to write a typical batch processing data pipeline using .NET for Apache Spark. We will show how a typical data processing job reads the source data and parses the data including dealing with any oddities the source files may have and then write the files out to a common format that other consumers of the data can use.

Imperfect Source Data

It is rare that when we are working with data sources, the files are in a perfect condition for processing; we often have to do some work to tidy the data, and in the example we will use in this chapter, this is as true as ever. We will be using some data published in the United Kingdom by Ofgem, the government regulator for gas and electricity markets. I found the data by browsing the UK government open data website. The files are useful for an example as they have several typical issues that we need to deal with when processing data.

The Source Data Files

The data files you need for this example can be downloaded from

- www.ofgem.gov.uk/system/files/docs/2017/05/12_mar_2017_ over_25k_spend_report.csv

- www.ofgem.gov.uk/system/files/docs/2017/05/01_apr_2017_ over_25k_spend_report.csv

- www.ofgem.gov.uk/system/files/docs/2017/06/02_may_2017_ over_25k_spend_report.csv

© Ed Elliott 2021
E. Elliott, *Introducing .NET for Apache Spark*, https://doi.org/10.1007/978-1-4842-6992-3_8

- www.ofgem.gov.uk/system/files/docs/2017/09/03_jun_2017_
  over_25k_spend_report.csv

- www.ofgem.gov.uk/system/files/docs/2017/09/04_jul_2017_
  over_25k_spend_report.csv

- www.ofgem.gov.uk/system/files/docs/2017/03/11_feb_2017_
  over_25k_spend_report.csv

If we examine the files, they are all CSV files, and the first few lines of the first file are in Listing 8-1.

Listing 8-1. The first few lines of 12_mar_2017_over_25k_spend_report.csv

```
Over 25k Expenditure Report,,,,,,,,,,
Date,Expense Type,Expense Area,Supplier,Reference, Amount ,,,,
March 2017,Building Rates,Corporate Services,CITY OF WESTMINSTER,58644,"
£1,807,657.66 ",,,,
March 2017,Building Rent,Corporate Services,CB RICHARD ELLIS,58332,"
£1,488,000.00 ",,,,
March 2017,Consultancy Fees, Ofgem ,PRICEWATERHOUSECOOPERS,58660,"
£187,870.80 ",,,,
```

The things to note are

1. The very first row is redundant; "Over 25k Expenditure Report,,,,,,," makes it look good in a spread sheet, but we need to read the second row with the actual column names in first.

2. There are several empty columns on each row, and each file has a different number of empty columns.

3. The date format is odd in that most of the files follow the pattern, month followed by year, but there is at least one file where the date format is month-year.

4. The Amount column includes padding, a £ sign, and a comma that do not make for easy converting into numerical values.

5. At the bottom of the file, not shown here, there are several empty rows.

Given these issues, when we read the files, we will need to do some additional work to make the data available for querying.

The Data Pipeline

We will create a data pipeline that will read in the source files and process them, one at a time, into a data lake, following these steps:

1. Read each source CSV file.

2. Remove the empty rows.

3. Use the column headers in the second row to give each column the correct name.

4. Remove the first row, which is a redundant header.

5. Convert the date into a usable date type.

6. Convert the amount into a usable numerical type.

7. Write the more usable data to the "Structured" area of the data lake.

8. Use the structured data to run some data validation rules.

9. Write the validated data to the "Curated" area of the data lake, partitioning the data by month and year.

10. Finally, take the curated data and write it in the delta format so that downstream processes and reports can consume the data.

Taking the source data and converting it into a known structure and then taking the data and curating and then publishing it are a typical pattern for working with data lakes. You may not use the same exact terms of source, structured, curated, and publish, but will likely have some variation of that. It may seem overly convoluted to use these different areas, but it allows us to be certain about what we have in the different areas. In Table 8-1, we look at what each of the areas is for and what we can expect when reading from it.

Table 8-1. *The different areas of a data lake*

Area	Description
Source	This is the raw source data in whatever format the source system gives the data in. It often cannot be relied upon to be used directly and requires processing to be useful
Structured	The raw data has been parsed into a common format; the data hasn't been validated but will be in a common format that is easier to read than the raw data. This area is, typically, the last time the data is stored in the same way as it was received in terms of column names and the same set of files as the data was received
Curated	In this area, the data has been validated and is ready to be used. We will normally see the data as a whole dataset with all the days, months, and years' worth of data as opposed to individual files
Publish	In this area, the data has normally been transformed into some sort of model, either dimensional modeling or data vault modeling. The data in this area will be consumed by a reporting tool or by advanced users

In this chapter, we will go through Listings 8-1 and 8-2 which are the full C# and F# versions of the pipeline, but we will explain each one, step by step. First, we will go through the C# version and then the F# version because the actual implementation is different. However, both achieve the same result, albeit slightly different due to differences in the languages. To follow through with these examples, you should download the files from the URLs already given and pass each file to your application using the command-line arguments.

When writing a data pipeline, several pieces of information are useful to pass into the application, the main one being the path to the data lake. In our example, we will reference a local folder and use that to test locally the pipeline. Still, the simplicity of Apache Spark means that we can write to a local file system when developing and testing and then pass in the path to a data lake in an Azure storage account or an AWS S3 bucket or Hadoop by changing the configuration and the path or URL to the data store, without having to change the code.

The C# Data Pipeline

We will now look at how to build this data pipeline in C#; because the F# implementation is slightly different, you can find the F# implementation later on in this chapter.

In Listing 8-2, we validate the arguments passed into our data pipeline, which should be the root path to the data lake, the source file, and also the year and month for which the file is.

Listing 8-2. Handling command-line arguments and getting setup

```
if (args.Length != 4)
{
    Console.WriteLine($"Error, incorrect args. Expecting 'Data Lake Path'
'file path' 'year' 'month', got: {args}");
    return;
}

var spark = SparkSession.Builder()
    .Config("spark.sql.sources.partitionOverwriteMode", "dynamic")
    .Config("spark.sql.extensions", "io.delta.sql.
DeltaSparkSessionExtension")
    .GetOrCreate();

var dataLakePath = args[0];
var sourceFile = args[1];
var year = args[2];
var month = args[3];

const string sourceSystem = "ofgem";
const string entity = "over25kexpenses";
```

Here, we do some basic validation on the arguments and then create a SparkSession. The SparkSession has an option set on it. The option is "spark.sql. sources.partitionOverwriteMode" which allows us to overwrite one specific partition rather than a complete table. This is useful because when we write to the curated part of the data lake, we will partition the data by year and month. Because each file is for a single month, if there were corrections, the file could be reissued; we want to process a month worth of data and overwrite anything that already exists for that month, but not the entire table. Without this option, we either could not overwrite a partition or would overwrite the whole table with the single partition.

We then do some simple argument parsing. You would probably add extra validation to the argument parsing or use a library to parse the arguments in a real-life application.

In Listing 8-3, we then create a method that orchestrates the pipeline; the steps are to read from the source file, write to the structured area and then to the curated area, and finally create the end result in the publish area.

Listing 8-3. Orchestrate the pipeline

```
private static void ProcessEntity(SparkSession spark, string sourceFile,
string dataLakePath, string sourceSystem, string entity, string year,
string month)
{
    var data = OfgemExpensesEntity.ReadFromSource(spark, sourceFile);

    OfgemExpensesEntity.WriteToStructured(data,$"{dataLakePath}/structured/
{sourceSystem}/{entity}/{year}/{month}");
    OfgemExpensesEntity.WriteToCurated(data, $"{dataLakePath}/curated/
{sourceSystem}/{entity}");

    OfgemExpensesEntity.WriteToPublish(spark, $"{dataLakePath}/curated/
{sourceSystem}/{entity}", $"{dataLakePath}/publish/{sourceSystem}/
{entity}");
}
```

The first thing we do is to read the source file and do some processing to get the correct columns and data types, so we can write to our preferred format in the structured area. Listing 8-4 shows reading the source file with the options we need for these files.

Listing 8-4. Reading the source file with the correct options

```
var dataFrame = spark.Read().Format("csv").Options(
    new Dictionary<string, string>()
    {
        {"inferSchema", "false"},
        {"header", "false" },
        {"encoding", "ISO-8859-1"},
        {"locale", "en-GB"},
        {"quote", "\""},
```

```
        {"ignoreLeadingWhiteSpace", "true"},
        {"ignoreTrailingWhiteSpace", "true"},
        {"dateFormat", "M y"}
    }
).Load(path);
```

This particular set of files has an extra header row which we need to skip. There are a couple of different approaches we could take here such as preprocessing the file to remove the extra header line before we go near it with Apache Spark. If I was writing this data pipeline in something like SSIS, then I probably would preprocess the file. In Listing 8-5, we show how to read in the whole file without using the column headers, by adding an index to the rows so we can filter out the first two rows but use the second row as the source of the column names. One potential issue with files like this is the data source providers adding or changing the columns, so we need to be careful relying on column ordering for names. It is normally more reliable to use the column names as they are supplied.

Listing 8-5. Using an index to skip over the additional header row and then using the column headers to name the columns and ignore any additional empty columns

```
private static readonly List<string> ColumnsToKeep = new List<string>()
    {
        "Date","Expense Type","Expense Area","Supplier","Reference",
        "Amount"
    };

//Add an index column with an increasing id for each row
var dataFrameWithId = dataFrame.WithColumn("index",
MonotonicallyIncreasingId());

//Pull out the column names
var header = dataFrameWithId.Filter(Col("index") == 1).Collect();

//filter out the header rows
var filtered = dataFrameWithId.Filter(Col("index") > 1).Drop("index");

var columnNames = new List<string>();
```

```
var headerRow = header.First();

for (int i = 0; i < headerRow.Values.Length; i++)
{
    if (headerRow[i] == null || !ColumnsToKeep.Contains(headerRow[i]))
    {
        Console.WriteLine($"DROPPING: _c{i}");
        filtered = filtered.Drop($"_c{i}");
    }
    else
    {
        columnNames.Add((headerRow[i] as string).Replace(" ", "_"));
    }
}

var output = filtered.ToDF(columnNames.ToArray());
```

The key here is that we use the function `MonotonicallyIncreasingId()`, which gives each row an index number we can use to then filter on later. We then drop any columns that we don't need and read the actual header row as an array of strings and take the filtered data that is just the actual data without the two header rows and call `filtered.ToDF`, passing in the names of the columns. This then gives us a `DataFrame` where we can now reference the columns using the column names.

Because the source data files include several rows that are all completely empty, in Listing 8-6, we filter out any rows that do not have either a supplier or a reference.

Listing 8-6. Filtering out rows that are empty

```
output = output.Filter(Col("Reference").IsNotNull() & Col("Supplier").
IsNotNull());
```

Now we have just the rows we are interested in. We will fix the Amount column. In Listing 8-7, we show how to remove the extra "£" and "," in the Amount column and convert the data to an actual numerical value, in this case, a float.

Listing 8-7. Turning the Amount string into a usable value

```
output = output.WithColumn("Amount", RegexpReplace(Col("Amount"), "[£,]", ""));
output = output.WithColumn("Amount", Col("Amount").Cast("float"));
```

The next column to deal with is the "Date" column, and they generally used one date format, but in some files, they changed to using a different date format, so we need to be able to cater to both possibilities. To handle this, in Listing 8-8, we take a copy of the existing column and then try and convert the date column. If the conversion results in all null values in the date column, then we attempt again using the alternate date format.

Listing 8-8. Dealing with multiple date formats

```
output = output.WithColumn("OriginalDate", Col("Date"));
output = output.WithColumn("Date", ToDate(Col("Date"), "MMMM yyyy"));

if (output.Filter(Col("Date").IsNull()).Count() == output.Count())
{
    Console.WriteLine("Trying alternate date format...");
    output = output.WithColumn("Date", ToDate(Col("OriginalDate"),
    "MMM-yy"));
}

output = output.Drop("OriginalDate");

return output;
```

Finally, we return the DataFrame which should have

- The correct column headers

- Any empty columns/rows removed

- The correct data types

Because we have gone through quite a lot of work to be able to read the files, it is often useful to save the files with the same raw data but in a format that can be read more easily. In Listing 8-9, we will write the file out as a parquet file to the "Structured" area of the data lake.

Listing 8-9. Writing out the raw data in a format that can be easily consumed

```
public static void WriteToStructured(DataFrame data, string path)
{
    data.Write().Mode("overwrite").Format("parquet").Save(path);
}
```

The path we are passing in is already scoped to the year and month, so we can overwrite whatever is there; otherwise, we would want to make sure we don't overwrite other months' data.

In the next phase, we will write to the "Curated" area, which means we need to perform some validations to ensure that we are only bringing in valid data.

In Listing 8-10, we will show the first validation to ensure that the schema we have matches the expected schema. This will validate that the correct columns exist and that the data types are correct. In this case, we just do an equality match to make sure the schemas are the same. In some systems, we might want to iterate the columns and check that we have x columns of y type as a minimum.

Listing 8-10. Validating the DataFrame schema

```
StructType _expectedSchema = new StructType(new List<StructField>()
    {
        new StructField("Date", new DateType()),
        new StructField("Expense_Type", new StringType()),
        new StructField("Expense_Area", new StringType()),
        new StructField("Supplier", new StringType()),
        new StructField("Reference", new StringType()),
        new StructField("Amount", new FloatType())
    });

if (data.Schema().Json != _expectedSchema.Json)
        {
            Console.WriteLine("Expected Schema Does NOT Match");
            Console.WriteLine("Actual Schema: " + data.Schema().
            SimpleString);
            Console.WriteLine("Expected Schema: " + _expectedSchema.
            SimpleString);
            ret = false;
        }
```

In Listing 8-11, we will show the next two checks, which are to validate that we have a value in every row of the date column and check that we have any data at all.

Listing 8-11. Validating the date column, and we have at least one row

```
if (data.Filter(Col("Date").IsNotNull()).Count() == 0)
        {
                Console.WriteLine("Date Parsing resulted in all NULL's");
                ret = false;
        }

        if (data.Count() == 0)
        {
                Console.WriteLine("DataFrame is empty");
                ret = false;
        }
```

The final check is more of a business rule, the data contains every supplier who charged more than £25,000 in a month, so we check that every amount is over £25,000. However, sometimes a single supplier provides different services. Each service could be less than £25,000, so we need to aggregate the Supplier column and Sum the Amount and then filter to see if there are any under £25,000. In Listing 8-12, we will show how to aggregate using the GroupBy function.

Listing 8-12. Using GroupBy and Sum to get a total amount for each supplier

```
var amountBySuppliers = data.GroupBy(Col("Supplier")).Sum("Amount")
    .Filter(Col("Sum(Amount)") < 25000);

if (amountBySuppliers.Count() > 0)
{
    Console.WriteLine("Amounts should only ever be over 25k");
    amountBySuppliers.Show();
    ret = false;
}
```

Once the data has been validated, we either write out the correct data or, if it fails validation, write it out as an error to investigate later. In Listing 8-13, we show how to write the data, but instead of a single file, we are writing with the rest of the data for the file and partitioning it by month and year. If anyone needs to read the data, they can read it from a single place, using filtering just to read the years and months they are interested in.

Listing 8-13. Writing the data into a common area using partitioning to keep it isolated from other years and months

```
if (ValidateEntity(data))
{
    data.WithColumn("year", Year(Col("Date")))
        .WithColumn("month", Month(Col("Date")))
        .Write()
        .PartitionBy("year", "month")
        .Mode("overwrite")
        .Parquet(path);
}
else
{
    Console.WriteLine("Validation Failed, writing failed file.");
    data.Write().Mode("overwrite").Parquet($"{path}-failed");
}
```

At this point, we have the raw data, we have the data in a more straightforward format for others to read in the "Structured" area, and we have the validated data in the "Curated" area. The final step is to write to the "Publish" area of the data lake.

The data in the "Publish" area will have two features. First, we will apply some data modeling in that instead of just having one big table, we will use a single fact table with a dimension table for each of the attributes that can be moved into a dimension. The second feature is that we will use the delta format for writing files. The delta format allows us to merge changes, so if we reprocess a file, then any updates will be merged into the data. The delta format gives us various interesting properties that are useful such as the ACID properties that we normally associate with an RDBMS. These ACID properties give us

- **Atomicity** – Either the write completes or doesn't complete, no partial completion.

- **Consistency** – The data is always in a valid state no matter when someone tries to read from it.

- **Isolation** – Multiple concurrent writes cannot corrupt the data.

- **Durability** – Once a write completes, then it will stay written, irrespective of a system failure.

Knowing that we could have multiple ETL jobs processing different files at the same time makes writing data pipelines significantly simpler.

The delta format achieves supporting the ACID properties using a transaction log file that it writes, and this transaction log further gives us the ability to read the data at a point in time, so we can read from the table but ask for the data as it appeared last week which is useful for production troubleshooting.

In Listing 8-14, we show the first part to the publish process; we read the data in from the "Curated" area. Although we have a single program for the entire process, this is often split into multiple jobs processing each step, so we show a complete divide between sections.

Listing 8-14. Read the data back in from the "Curated" area

```
var data = spark.Read().Parquet(rootPath);
```

Now we have the data we will take the dimension like attributes from the data and split them into their own delta tables. The end goal for this is to have a single fact delta table with the values such as date and amount, and the attributes such as "Supplier" and "Expense Type" will be in their own delta tables. In Listing 8-15, we will read the "Supplier" column and create a hash of the supplier name which will be the key that we can use to join back to the main fact delta table. The reason for using a hash of the supplier name rather than an incrementing key is, if we wish, we could load any dimensions and facts in parallel jobs. If the data had to exist in the dimension before loading the fact delta table, then we would need to be stricter in the order of the processing. Once we have added the key column to the supplier, if this is the first time we have written to the delta table, then we will create a new table. If it isn't the first ever file we are processing, then we will join to the existing data using a "left_anti" join which means only give rows on the left which do not exist on the right. We then insert the new rows into the delta table. It should be evident how using the delta format for writing really starts to make processing in a data lake similar to processing in an RDBMS or SQL database.

Listing 8-15. Storing each supplier in a dimension delta table

```
var suppliers = data.Select(Col("Supplier")).Distinct()
                        .WithColumn("supplier_hash",
Hash(Col("Supplier"))));

var supplierPublishPath = $"{publishPath}-suppliers";

if (!Directory.Exists(supplierPublishPath))
{
    suppliers.Write().Format("delta").Save(supplierPublishPath);
}
else
{
    var existingSuppliers = spark.Read().Format("delta").
    Load(supplierPublishPath);
    var newSuppliers = suppliers.Join(existingSuppliers,
existingSuppliers["Supplier"] == suppliers["Supplier"], "left_anti");
    newSuppliers.Write().Mode(SaveMode.Append).Format("delta").
Save(supplierPublishPath);
}
```

In Listing 8-16, we do the same thing but with the "Expense Type" column; we move the data into its own dimension delta table.

Listing 8-16. Move the "Expense Type" column into its own dimension delta table

```
var expenseTypePublishPath = $"{publishPath}-expense-type";

var expenseType = data.Select(Col("Expense_Type")).Distinct().
WithColumn("expense_type_hash", Hash(Col("Expense_Type")));
```

```
if (!Directory.Exists(expenseTypePublishPath))
{
    expenseType.Write().Format("delta").Save(expenseTypePublishPath);
}
else
{
    var existingExpenseType = spark.Read().Format("delta").
    Load(expenseTypePublishPath);
    var newExpenseType = expenseType.Join(existingExpenseType,
    existingExpenseType["Expense_Type"] == expenseType["Expense_Type"],
    "left_anti");
    newExpenseType.Write().Mode(SaveMode.Append).Format("delta").
    Save(expenseTypePublishPath);
}

data = data.WithColumn("Expense_Type", Hash(Col("Expense_Type"))).
WithColumn("Supplier", Hash(Col("Supplier")));
```

In Listing 8-17, which is the final part of the publish phase, if this is the first ever file we have processed, then we can just write the data as delta format; if the data already exists, then we will merge the data together which either updates any existing amounts if there was an update or inserts new rows.

Listing 8-17. Using a merge to write into the existing data

```
if (!Directory.Exists(publishPath))
{
    data.Write().Format("delta").Save(publishPath);
}
else
{
    var target = DeltaTable.ForPath(publishPath).Alias("target");
    target.Merge(
        data.Alias("source"), "source.Date = target.Date AND source.
        Expense_Type = target.Expense_Type AND source.Expense_Area =
        target.Expense_Area AND source.Supplier = target.supplier AND
        source.Reference = target.Reference"
        ).WhenMatched("source.Amount != target.Amount")
```

157

```
        .Update(new Dictionary<string, Column>(){{"Amount",
        data["Amount"]}}
).WhenNotMatched()
        .InsertAll()
.Execute();
}
```

The merge statement itself is interesting. It allows us to merge a source and a target DataFrame by specifying which columns should match; if we find matches, then we can update or, optionally, provide an additional filter and then update as we have done here: `WhenMatched("source.Amount != target.Amount")`. We can choose what to do if the merge criteria determine that a row doesn't exist; here, we just want to insert all the rows, but we could be more selective about which columns we insert. Finally, to run the merge statement, we need to call `Execute`.

One thing to note is that, for now, the `Merge` statement is a bit of a mix of code and SQL, and to make the SQL explicit about which is the source and target, I use alias on both the `DeltaTable` and the `DataFrame` to ensure that there is no confusion about what we are comparing and when.

It is also possible to completely write the merge statement using SQL by adding another option to the SparkSession, "spark.sql.extensions", which should be set to "io.delta.sql.DeltaSparkSessionExtension". If we used this option, we could replace our code with a SQL merge statement.

The F# Data Pipeline

In Listing 8-18, we validate the arguments passed into our data pipeline, which should be the root path to the data lake, the source file, and also the year and month for which the file is.

Listing 8-18. Handling command-line arguments and getting setup

```
let args = match argv with
            | [|dataLakePath; path; year; month|] -> {dataLakePath = argv.
              [0]; path = argv.[1]; year = argv.[2]; month = argv.[3];
              success = true}
```

```
            | _ -> {success = false; dataLakePath= ""; path = ""; year =
              ""; month = "";}

match args.success with
    | false ->
        printfn "Error, incorrect args. Expecting 'Data Lake Path' 'file
        path' 'year' 'month', got: %A" argv
        -1

    | true ->
            let spark = SparkSession.Builder().Config("spark.sql.sources.
            partitionOverwriteMode", "dynamic").GetOrCreate()
```

Here, we do some basic validation on the arguments and then create a
SparkSession. The SparkSession has an option set on it. The option is "spark.sql.
sources.partitionOverwriteMode" which allows us to overwrite one specific partition
rather than a complete table. This is useful because when we write to the curated part
of the data lake we will partition the data by year and month and because each file is for
a single month, if there were corrections, the file could be reissued; we want to process
a month worth of data and overwrite anything that already exists for that month, but
not the entire table. Without this option, we either could not overwrite a partition or
overwrite the whole table with the single partition.

We then do some simple argument parsing. You would probably add extra validation
to the argument parsing or use a library to parse the arguments in a real-life application.

In Listing 8-19, we then create a method that orchestrates the pipeline; the steps are
read from the source file, write to the structured area and then to the curated area, and
finally create the end result in the publish area.

Listing 8-19. Orchestrate the pipeline

```
let data = getData(spark, args.path)

                writeToStructured (data, (sprintf "%s/structured/%s/%s/%s/%s"
                args.dataLakePath "ofgem" "over25kexpenses" args.year args.
                month))
                match validateEntity data with
                  | false -> writeToFailed(data, (sprintf "%s/failed/%s/%s"
                    args.dataLakePath "ofgem" "over25kexpenses"))
```

```
                          -2
            | true -> writeToCurated(data, (sprintf "%s/curated/%s/%s"
                args.dataLakePath "ofgem" "over25kexpenses"))
                        writeToPublished(spark,  (sprintf "%s/
                        curated/%s/%s" args.dataLakePath "ofgem"
                        "over25kexpenses"), (sprintf "%s/publish/%s/%s"
                        args.dataLakePath "ofgem" "over25kexpenses"))
                        0
```

The first thing we do is to read the source file and do some processing to get the correct columns and data types, so we can write to our preferred format in the structured area. Listing 8-20 shows reading the source file with the options we need for these files.

Listing 8-20. Reading the source file with the correct options

```
let getData(spark:SparkSession, path:string) =
    let readOptions =
        let options = [
            ("inferSchema","true")
            ("header","false")
            ("encoding","ISO-8859-1")
            ("locale","en-GB")
            ("quote","\"")
            ("ignoreLeadingWhiteSpace","true")
            ("ignoreTrailingWhiteSpace","true")
            ("dateFormat","M y")
        ]
        System.Linq.Enumerable.ToDictionary(options, fst, snd)

    spark.Read().Format("csv").Options(readOptions).Load(path)
        |> fun data -> data.WithColumn("index", Functions.
           MonotonicallyIncreasingId())
        |> dropIgnoredColumns
        |> fixColumnHeaders
        |> filterOutEmptyRows
        |> fixDateColumn
        |> fixAmountColumn
```

This particular set of files has an extra header row which we need to skip. There are a couple of different approaches we could take here such as preprocessing the file to remove the extra header line before we go near it with Apache Spark. If I was writing this data pipeline in something like SSIS, then I probably would preprocess the file. In Listing 8-21, we show how to read in the whole file without using the column headers, by adding an index to the rows so we can filter out the first two rows but use the second row as the source of the column names. One potential issue with files like this is the data source providers adding or changing the columns, so we need to be careful relying on column ordering for names. It is normally more reliable to use the column names as they are supplied.

Listing 8-21. Using an index to skip over the additional header row and then using the column headers to name the columns and ignore any additional empty columns

```
let dropIgnoredColumns (dataFrameToDropColumns:DataFrame) : DataFrame =

    let header = dataFrameToDropColumns.Filter(Functions.Col("index").
    EqualTo(1)).Collect()

    let shouldDropColumn (_:int, data:obj) =
        match data with
            | null -> true
            | _ -> match data.ToString() with
                        | "Date" -> false
                        | "Expense Type" -> false
                        | "Expense Area" -> false
                        | "Supplier" -> false
                        | "Reference" -> false
                        | "Amount" -> false
                        | "index" -> false
                        | null -> true
                        | _ -> true

    let dropColumns =
        let headerRow = header |> Seq.cast<Row> |> Seq.head
        headerRow
```

```
            |> fun row -> row.Values
            |> Seq.indexed
            |> Seq.filter shouldDropColumn
            |> Seq.map fst
            |> Seq.map(fun i -> "_c" + i.ToString())
            |> Seq.toArray

        dataFrameToDropColumns.Drop dropColumns

let fixColumnHeaders (dataFrame:DataFrame) : DataFrame =
    let header = getHeaderRow dataFrame
                    |> convertHeaderRowIntoArrayOfNames

    dataFrame.Filter(Functions.Col("index").Gt(1)).Drop("index").
    ToDF(header)
```

The key here is that we use the function `MonotonicallyIncreasingId()`, which gives each row an index number we can use to then filter on later. We then drop any columns that we don't need and read the actual header row as an array of strings and take the filtered data that is just the actual data without the two header rows and call `ToDF(header)` where the header is the new set of column names. This then gives us a `DataFrame` where we can now reference the columns using the column names.

Because the source data files include several rows that are all completely empty, in Listing 8-22, we filter out any rows that do not have either a supplier or a reference.

Listing 8-22. Filtering out rows that are empty

```
let filterOutEmptyRows (dataFrame:DataFrame) : DataFrame =
    dataFrame.Filter(Functions.Col("Reference").IsNotNull()).
Filter(Functions.Col("Supplier").IsNotNull())
```

Now we have just the rows we are interested in. We will fix the Amount column. In Listing 8-23, we show how to remove the extra "£" and "," in the Amount column and convert the data to an actual numerical value, in this case, a float.

Listing 8-23. Turning the Amount string into a usable value

```
let fixAmountColumn (dataFrame:DataFrame) : DataFrame =
    dataFrame.WithColumn("Amount", Functions.RegexpReplace(Functions.
    Col("Amount"), "[£,]", ""))
    |> fun d -> d.WithColumn("Amount", Functions.Col("Amount").
       Cast("float"))
```

The next column to deal with is the "Date" column, and they generally used one date format, but in some files, they changed to using a different date format, so we need to be able to cater to both possibilities. To handle this, in Listing 8-24, we take a copy of the existing column and then try and convert the date column. If the conversion results in all null values in the date column, then we attempt again using the alternate date format.

Listing 8-24. Dealing with multiple date formats

```
let fixDateColumn (dataFrame:DataFrame) : DataFrame =
    dataFrame.WithColumn("__Date", Functions.Col("Date"))
    |> fun d -> d.WithColumn("Date", Functions.ToDate(Functions.
       Col("Date"), "MMMM yyyy"))
    |> fun d-> match d.Filter(Functions.Col("Date").IsNotNull()).Count()
       with
                  | 0L -> d.WithColumn("Date", Functions.
                    ToDate(Functions.Col("__Date"), "MMM-yy"))
                  | _ -> d
    |> fun d -> d.Drop("__Date")
```

Finally, we return the DataFrame which should have

- The correct column headers

- Any empty columns/rows removed

- The correct data types

Because we have gone through quite a lot of work to be able to read the files, it is often useful to save the files with the same raw data but in a format that can be read more easily. In Listing 8-25, we will write the file out as a parquet file to the "Structured" area of the data lake.

Listing 8-25. Writing out the raw data in a format that can be easily consumed

```
let writeToStructured(dataFrame:DataFrame, path:string) : unit =
    dataFrame.Write().Mode("overwrite").Format("parquet").Save(path)
```

The path we are passing in is already scoped to the year and month, so we can overwrite whatever is there; otherwise, we would want to make sure we don't overwrite other months' data.

In the next phase, we will write to the "Curated" area, which means we need to perform some validations to ensure that we are only bringing in valid data.

In Listing 8-26, we will show the first validation to ensure that the schema we have matches the expected schema. This will validate that the correct columns exist and that the data types are correct. In this case, we just do an equality match to make sure the schemas are the same. In some systems, we might want to iterate the columns and check that we have x columns of y type as a minimum.

Listing 8-26. Validating the DataFrame schema

```
let expectedSchema = StructType(
                            [|
                                StructField("Date", DateType())
                                StructField("Expense_Type",
                                StringType())
                                StructField("Expense_Area",
                                StringType())
                                StructField("Supplier",
                                StringType())
                                StructField("Reference",
                                StringType())
                                StructField("Amount", FloatType())
                            |]
                        )

let validateSchema (dataFrame:DataFrame) = dataFrame.Schema().Json =
expectedSchema.Json
```

In Listing 8-27, we will show the next two checks, which are to validate that we have a value in every row of the date column and check that we have any data at all.

Listing 8-27. Validating the date column, and we have at least one row

```
let validateHaveSomeNonNulls (dataFrame:DataFrame) = dataFrame.
Filter(Functions.Col("Date").IsNotNull()).Count() > 0L
```

The final check is more of a business rule; the data contains every supplier who charged more than £25,000 in a month, so we check that every amount is over £25,000. However, sometimes, a single supplier provides different services. Each service could be less than £25,000, so we need to aggregate the Supplier column and Sum the Amount and then filter to see if there are any under £25,000. In Listing 8-28, we will show how to aggregate using the GroupBy function.

Listing 8-28. Using GroupBy and Sum to get a total amount for each supplier

```
let validateAmountsPerSupplierGreater25K (dataFrame:DataFrame) = dataFrame.
GroupBy(Functions.Col("Supplier")).Sum("Amount").Filter(Functions.
Col("Sum(Amount)").Lt(25000)).Count() = 0L
```

Once the data has been validated, we either write out the correct data or, if it fails validation, write it out as an error to investigate later. In Listing 8-29, we show how to write the data, but instead of a single file, we are writing with the rest of the data for the file and partitioning it by month and year. If anyone needs to read the data, they can read it from a single place, using filtering just to read the years and months they are interested in.

Listing 8-29. Writing the data into a common area using partitioning to keep it isolated from other years and months

```
let writeToCurated (dataFrame:DataFrame, path:string) : unit =
    dataFrame.WithColumn("year", Functions.Year(Functions.Col("Date")))
    |> fun data -> data.WithColumn("month", Functions.Month(Functions.
Col("Date")))
    |> fun data -> data.Write().PartitionBy("year", "month").
Mode("overwrite").Parquet(path);
```

At this point, we have the raw data, we have the data in a more straightforward format for others to read in the "Structured" area, and we have the validated data in the "Curated" area. The final step is to write to the "Publish" area of the data lake.

The data in the "Publish" area will have two features. First, we will apply some data modeling in that instead of just having one big table, we will use a single fact table with a dimension table for each of the attributes that can be moved into a dimension. The second feature is that we will use the delta format for writing files. The delta format allows us to merge changes, so if we reprocess a file, then any updates will be merged into the data. The delta format gives us various interesting properties that are useful such as the ACID properties that we normally associate with an RDBMS. These ACID properties give us

- **Atomicity** – Either the write completes or doesn't complete, no partial completion.

- **Consistency** – The data is always in a valid state no matter when someone tries to read from it.

- **Isolation** – Multiple concurrent writes cannot corrupt the data.

- **Durability** – Once a write completes, then it will stay written, irrespective of a system failure.

Knowing that we could have multiple ETL jobs processing different files at the same time makes writing data pipelines significantly simpler.

The delta format achieves supporting the ACID properties using a transaction log file that it writes, and this transaction log further gives us the ability to read the data at a point in time, so we can read from the table but ask for the data as it appeared last week which is useful for production troubleshooting.

In Listing 8-30, we show the first part to the publish process; we read the data in from the "Curated" area. Although we have a single program for the entire process, this is often split into multiple jobs processing each step, so we show a complete divide between sections.

Listing 8-30. Read the data back in from the "Curated" area

```
let data = spark.Read().Parquet(source)
```

Now we have the data we will take the dimension like attributes from the data and split them into their own delta tables. The end goal for this is to have a single fact delta table with the values such as date and amount, and the attributes such as "Supplier" and "Expense Type" will be in their own delta tables. In Listing 8-31, we will read the "Supplier" column and create a hash of the supplier name which will be the key that we can use to join back to the main fact delta table. The reason for using a hash of the supplier name rather than an incrementing key is, if we wish, we could load any dimensions and facts in parallel jobs. If the data had to exist in the dimension before loading the fact delta table, then we would need to be stricter in the order of the processing. Once we have added the key column to the supplier, if this is the first time we have written to the delta table, then we will create a new table. If it isn't the first ever file we are processing, then we will join to the existing data using a "left_anti" join which means only give rows on the left which do not exist on the right. We then insert the new rows into the delta table. It should be evident how using the delta format for writing really starts to make processing in a data lake similar to processing in an RDBMS or SQL database.

Listing 8-31. Storing each supplier in a dimension delta table

```
let saveSuppliers (spark: SparkSession, dataFrame:DataFrame, source:string,
target:string) =

    let suppliers = dataFrame.Select(Functions.Col("Supplier")).Distinct()

    match Directory.Exists(sprintf "%s-suppliers" target) with
        | true -> let existingSuppliers = spark.Read().Format("delta").
          Load(sprintf "%s-suppliers" target)
                    existingSuppliers.Join(suppliers, existingSuppliers.
                    Col("Supplier").EqualTo(suppliers.Col("Supplier")),
                    "left_anti")
                      |> fun newSuppliers -> newSuppliers.
WithColumn("Supplier_Hash", Functions.Hash(Functions.Col("Supplier"))).
Write().Mode(SaveMode.Append).Format("delta").Save(sprintf "%s-suppliers"
target)
        | false -> suppliers.WithColumn("Supplier_Hash", Functions.
          Hash(Functions.Col("Supplier"))).Write().Format("delta").
          Save(sprintf "%s-suppliers" target)
```

In Listing 8-32, we do the same thing but with the "Expense Type" column; we move the data into its own dimension delta table.

Listing 8-32. Move the "Expense Type" column into its own dimension delta table

```
let saveExpenseType (spark: SparkSession, dataFrame:DataFrame,
source:string, target:string) =

    let expenseType = dataFrame.Select(Functions.Col("Expense_Type")).
    Distinct()

    match Directory.Exists(sprintf "%s-expense-type" target) with
        | true -> let existingExpenseType = spark.Read().Format("delta").
          Load(sprintf "%s-expense-type" target)
                    existingExpenseType.Join(expenseType,
                    existingExpenseType.Col("Expense_Type").
                    EqualTo(expenseType.Col("Expense_Type")), "left_anti")
                        |> fun newExpenseType -> newExpenseType.
                        WithColumn("Expense_Type_Hash", Functions.
                        Hash(Functions.Col("Expense_Type"))).Write().
                        Mode(SaveMode.Append).Format("delta").Save(sprintf
                        "%s-expense-type" target)
        | false -> expenseType.WithColumn("Expense_Type_Hash", Functions.
          Hash(Functions.Col("Expense_Type"))).Write().Mode("overwrite").
          Format("delta").Save(sprintf "%s-expense-type" target)
```

In Listing 8-33, which is the final part of the publish phase, if this is the first ever file we have processed, then we can just write the data as delta format; if the data already exists, then we will merge the data together which either updates any existing amounts if there was an update or inserts new rows.

Listing 8-33. Using a merge to write into the existing data

```
let writeExpenses (dataFrame:DataFrame, target:string) =

    let data = dataFrame.WithColumn("Expense_Type_Hash", Functions.
    Hash(Functions.Col("Expense_Type"))).Drop("Expense_Type")
```

```
            |> fun data -> data.WithColumn("Supplier_Hash", Functions.
                Hash(Functions.Col("Supplier"))).Drop("Supplier").
                Alias("source")

    match Directory.Exists(target) with
        | false -> data.Write().Format("delta").Save(target)
        | true ->  DeltaTable.ForPath(target).Alias("target").Merge(data,
            "source.Date = target.Date AND source.Expense_Type_Hash = target.
            Expense_Type_Hash AND source.Expense_Area = target.Expense_Area
            AND source.Supplier_Hash = target.Supplier_Hash AND source.
            Reference = target.Reference")
                        |> fun merge -> let options = System.Linq.Enumerable.
                            ToDictionary(["Amount", data.["Amount"]], fst, snd)
                                    merge.WhenMatched("source.Amount !=
                                    target.Amount").Update(options)
                        |> fun merge -> merge.WhenNotMatched().InsertAll()
                        |> fun merge -> merge.Execute()
```

The merge statement itself is interesting. It allows us to merge a source and a target DataFrame by specifying which columns should match; if we find matches, then we can update or, optionally, provide an additional filter and then update as we have done here: `WhenMatched("source.Amount != target.Amount")`. We can choose what to do if the merge criteria determine that a row doesn't exist; here, we just want to insert all the rows, but we could be more selective about which columns we insert. Finally, to run the merge statement, we need to call `Execute`.

One thing to note is that, for now, the `Merge` statement is a bit of a mix of code and SQL, and to make the SQL explicit about which is the source and target, I use alias on both the `DeltaTable` and the `DataFrame` to ensure that there is no confusion about what we are comparing and when.

It is also possible to completely write the merge statement using SQL by adding another option to the SparkSession, "spark.sql.extensions", which should be set to "io. delta.sql.DeltaSparkSessionExtension". If we used this option, we could replace our code with a SQL merge statement.

Summary

Writing data pipelines in Apache Spark, either using .NET for Apache Spark or in Python, Scala, and so on, is typically a matter of breaking the processing up into a series of smaller steps and validating the data at each stage. Pretty much the only constant when receiving data is that the data will be wrong at some point, so it is a matter of making sure that you are able to debug your data pipelines efficiently and understand where and when they fail.

In this chapter, we have shown how to read from a data file which contains a number of challenges and process those individual data files into a complete dataset which has been validated and is ready for the business to consume.

CHAPTER 9

Structured Streaming

In this chapter, we will look at an example of how to create a streaming application. Apache Spark's structured streaming API allows you to use the DataFrame API to express your Apache Spark job. Instead of working with static datasets, you work with micro-batches of data using the scalable, fault-tolerant stream processing engine built on Apache Spark.

The application we will create will do two things. Firstly, it will examine every message for a specific condition and allow our application to raise an alert, and secondly it will gather all the data received within a 5-minute window, aggregate the data, and save the data so that it can be shown in a dashboard.

Our Streaming Example

In this chapter's example, we will use an Apache Kafka topic to read changes from a Microsoft SQL Server using Change Data Capture (CDC) using the Debezium connector. Aside from Microsoft SQL Server, these are all open source products. Configuring Microsoft SQL Server, Apache Kafka, and Debezium is outside of the scope of this chapter, but we will explain how to parse the Apache Kafka messages.

The overview of the entire process is that

1. Data is written to a SQL Server database.

2. SQL Server's Change Data Capture feature, generates change data.

3. Debezium reads the changes and posts to an Apache Kafka topic.

4. Our application reads from the Apache Kafka topic and processes the data.

171

© Ed Elliott 2021
E. Elliott, *Introducing .NET for Apache Spark*, https://doi.org/10.1007/978-1-4842-6992-3_9

It should be noted that although we are going to use Apache Kafka, there are a number of different sources that Apache Spark can stream from, and although the connection details and parsing are different, the processing within Apache Spark is the same.

Setting Up the Example

To run the example yourself, you will need an instance of SQL Server that supports Change Data Capture, an instance of Apache Kafka running with Kafka Connect, and the Debezium connector for SQL Server. In Listing 9-1, we show the table in SQL Server that we will use as the source table.

For more information on configuring SQL Server Change Data Capture, see https://docs.microsoft.com/en-us/sql/relational-databases/track-changes/ about-change-data-capture-sql-server. For Kafka and Debezium, see *https:// debezium.io/documentation/reference/connectors/sqlserver.html.*

Listing 9-1. The source SQL Server table

```
CREATE TABLE dbo.SalesOrderItems
(
    Order_Item_ID      INT IDENTITY(1,1) NOT NULL PRIMARY KEY,
    Order_ID           INT NOT NULL,
    Product_ID         INT NOT NULL,
    Amount INT         NOT NULL,
    Price_Sold         FLOAT NOT NULL,
    Margin FLOAT       NOT NULL
    Order_Date         DATETIME NOT NULL
)
```

Once the table has been created and CDC enabled on the database and the table, we can create a connection from the Debezium connector for SQL Server. Debezium will create the topic for us. In this example, the topic will be called "sql.dbo.SalesOrderItems" because we configure the name of the database in the Debezium connector configuration as "sql", and then the schema and name of the table are added to build up the complete topic name. If you created a connection to a separate table, then the name of that table would be used to create the topic.

When the data is written to the SQL Server table, Debezium reads any changes and creates a JSON message that is posted to Apache Kafka. In Listing 9-2, we show an example message which we will need to parse later on using the DataFrame API.

Listing 9-2. A sample Apache Kafka message from the Debezium SQL Server connector. The schema section has been removed to keep the size of the listing reasonable

```
{
    "schema": {
        "type": "struct",

        ],
        "optional": false,
        "name": "sql.dbo.SalesOrderItems.Envelope"
    },
    "payload": {
        "before": null,
        "after": {
            "Order_ID": 1,
            "Order_Item_ID": 737,
            "Product_ID": 123,
            "Amount": 10,
            "Price_Sold": 1000.23,
            "Margin": 0.99
        },
        "source": {
            "version": "1.3.0.Final",
            "connector": "sqlserver",
            "name": "sql",
            "ts_ms": 1602915585290,
            "snapshot": "false",
            "db": "Transactions",
            "schema": "dbo",
            "table": "SalesOrderItems",
            "change_lsn": "0000002c:00000c60:0003",
```

```
      "commit_lsn": "0000002c:00000c60:0004",
      "event_serial_no": 1
    },
    "op": "c",
    "ts_ms": 1602915587594,
    "transaction": null
  }
}
```

The JSON message is made up of a schema, the payload, which contains before and after data, and the source information such as the transaction time. In this example, the payload section has an empty "before" object because this was an insert. If it had been an update, then there would have been data in the before section.

The Streaming Application

In Listings 9-3 and 9-4, we will show how to use a SparkSession to create a DataFrame from the Apache Kafka topic. We are hard-coding the connection and topic information, but you would likely read it from the command-line arguments or a config file.

Listing 9-3. Creating a DataFrame from an Apache Kafka topic in C#

```
var spark = SparkSession.Builder().GetOrCreate();

var rawDataFrame = spark.ReadStream().Format("kafka")
    .Option("kafka.bootstrap.servers", "localhost:9092")
    .Option("subscribe", "sql.dbo.SalesOrderItems")
    .Option("startingOffset", "earliest").Load();
```

Listing 9-4. Creating a DataFrame from an Apache Kafka topic in F#

```
let spark = SparkSession.Builder().GetOrCreate();

let rawDataFrame = spark.ReadStream()
                |> fun stream -> stream.Format("kafka")
                |> fun stream -> stream.Option("kafka.bootstrap.
                    servers", "localhost:9092")
```

```
|> fun stream -> stream.Option("subscribe", "sql.dbo.
   SalesOrderItems")
|> fun stream -> stream.Option("startingOffset",
   "earliest")
|> fun stream -> stream.Load()
```

The "startingOffset" option we passed in determines the start point when a query is started. For a full list of available options and their descriptions, see `https://spark.apache.org/docs/latest/structured-streaming-kafka-integration.html`.

The "rawDataFrame" we have created here will read from the Apache Kafka topic so that the correct columns on the DataFrame can be built but will not contain any data at this point, and, in fact, if we tried to do a `rawDataFrame.Show()`, we would get an error message that says `"Queries with streaming sources must be executed with writeStream.start()"`. We can do a PrintSchema() however, and the DataFrame schema should look like this:

```
root
 |-- key: binary (nullable = true)
 |-- value: binary (nullable = true)
 |-- topic: string (nullable = true)
 |-- partition: integer (nullable = true)
 |-- offset: long (nullable = true)
 |-- timestamp: timestamp (nullable = true)
 |-- timestampType: integer (nullable = true)
```

The data is in the "value" column and is a binary representation of the JSON message we saw in Listing 9-3.

Shredding the JSON Document

Because the JSON is in a DataFrame column and we have the data already in a DataFrame, we want to use Apache Spark to shred the JSON document and retrieve the actual columns that we want. We will need to supply a schema that allows Apache Spark to read the JSON. In Listings 9-5 and 9-6, we show how to create the StructType schema definition.

Listing 9-5. Creating a StructType schema definition in C#

```
var jsonSchema = new StructType(
    new List<StructField>
    {
        new StructField("schema", new StringType()),
        new StructField("payload", new StructType(
            new List<StructField>
            {
                new StructField("after", new StructType(
                    new List<StructField>
                    {
                        new StructField("Order_ID", new IntegerType()),
                        new StructField("Product_ID", new IntegerType()),
                        new StructField("Amount", new IntegerType()),
                        new StructField("Price_Sold", new FloatType()),
                        new StructField("Margin", new FloatType())
                    })),
                new StructField("source", new StructType(new
                List<StructField>
                {
                    new StructField("version", new StringType()),
                    new StructField("ts_ms", new LongType())
                }))
            }))
    }
);
```

Listing 9-6. Creating a StructType schema definition in F#

```fsharp
let messageSchema() =
    StructType(
        [|
            StructField("schema", StringType())
            StructField("payload", StructType(
                [|
                    StructField("after", StructType(
                        [|
                            StructField("Order_ID", IntegerType())
                            StructField("Product_ID", IntegerType())
                            StructField("Amount", IntegerType())
                            StructField("Price_Sold", FloatType())
                            StructField("Margin", FloatType())
                        |]
                    ))
                    StructField("source", StructType(
                        [|
                            StructField("version", StringType())
                            StructField("ts_ms", LongType())
                        |]
                    ))
                |]
            ))
        |]
    )
```

The important thing to note is that you only need to supply the details for the parts of the document you are interested in. For example, the schema section of the document is useful for us to understand the document, but we can't use it from Apache Spark as we need the schema to read the schema, so in our StructType schema, we mark the whole section as a StringType so it is stored as a string which we can choose to read or not. For the parts of the document we are interested in, namely, the time of the transaction "ts_ms" and the order details, we do need to provide a specific schema. It should be noted that if you miss a column off, then it will just be ignored by Apache Spark, but if you provide an incorrect data type, then the entire row will be null, even if other values in the document have the correct data type.

Creating a DataFrame

In Listings 9-7 and 9-8, we will take the DataFrame we created that points to the Apache Kafka topic as well as our StructType schema and create a DataFrame where Apache Spark shreds the JSON document into actual DataFrame Columns we can work with.

Listing 9-7. Shredding the JSON document into DataFrame Columns in C#

```
var parsedDataFrame = rawDataFrame
    .SelectExpr("CAST(value as string) as value")
    .WithColumn("new", FromJson(Col("value"), messageSchema.Json))
    .Select("value", "new.payload.after.*", "new.payload.source.*")
    .WithColumn("timestamp", Col("ts_ms").Divide(1000).Cast("timestamp"));
```

Listing 9-8. Shredding the JSON document into DataFrame Columns in F#

```
let parsedDataFrame = rawDataFrame
                    |> fun dataFrame -> dataFrame.
                        SelectExpr("CAST(value as string) as value")
                    |> fun dataFrame -> dataFrame.WithColumn("new",
                        Functions.FromJson(Functions.Col("value"),
                        messageSchema().Json))
                    |> fun dataFrame -> dataFrame.Select("new.payload.
                        after.*", "new.payload.source.*")
                    |> fun dataFrame -> dataFrame.
                        WithColumn("timestamp", Functions.Col("ts_ms").
                        Divide(1000).Cast("timestamp"))
```

The points to note here are that we take the binary "value" column and cast it to a string, and then we use the FromJson function, combined with our schema. The FromJson Apache Spark function will create a column for every property in our schema, and we select the data using the JSON path "new.payload.after.*" which will give us one column for every one of the types specified in the schema under the "after" object, and the names will be the property name such as "Order_ID" and "Margin".

The timestamp provided by Microsoft Change Data Capture needs to be divided by 1000 in order that we can convert it to a timestamp with the correct date and time in.

At this point, we are still working with the original DataFrame that was created from the Apache Kafka topic. Without any data, we have not yet started streaming any data from anywhere.

Starting a Stream

The next thing we will do, in Listings 9-9 and 9-10, is to start a stream and use the ForeachBatch method, which will run exactly once for every micro-batch that Apache Spark structured streaming provides our application. We will use this micro-batch to examine every row and trigger an alert if an item is sold that matches a specific condition.

Listing 9-9. Using ForeachBatch to process each micro-batch looking for specific conditions in C#

```
var operationalAlerts = parsedDataFrame
    .WriteStream()
    .Format("console")
    .ForeachBatch((df, id) => HandleStream(df, id))
    .Start();

private static void HandleStream(DataFrame df, in long batchId)
{
    var tooLowMargin = df.Filter(Col("Margin").Lt(0.10));

    if (tooLowMargin.Count() > 0)
    {
        tooLowMargin.Show();
        Console.WriteLine("Trigger Ops Alert Here");
    }
}
```

```
private static void HandleStream(DataFrame df, in long batchId)
{
    var tooLowMargin = df.Filter(Col("Margin").Lt(0.10));

    if (tooLowMargin.Count() > 0)
    {
        tooLowMargin.Show();
        Console.WriteLine("Trigger Ops Alert Here");
    }
}
```

Listing 9-10. Using ForeachBatch to process each micro-batch looking for specific conditions in F#

```
let handleStream (dataFrame:DataFrame, _) : unit  =
    dataFrame.Filter
    (Functions.Col("Margin").Lt(0.10))
    |> fun failedRows -> match failedRows.Count() with
        | 0L -> printfn "We had no failing rows"
            failedRows.Show()
        | _ -> printfn "Trigger Ops Alert Here"

let operationalAlerts = parsedDataFrame.WithWatermark("timestamp",
"30 seconds")
    |> fun dataFrame -> dataFrame.WriteStream()
    |> fun stream -> stream.ForeachBatch(fun dataFrame batchId -> handleStr
        eam(dataFrame,batchId))
    |> fun stream -> stream.Start()
```

In `HandleStream`, we show that we can start to use the DataFrame API with the familiar methods we expect, such as Filter and Show, to build up our applications as we would when writing batch mode applications.

Two things to note here are the `WithWatermark` function and the `WriteStream` function. The `WithWatermark` function allows Apache Spark to make sure that late delivered messages are never dropped. In this system, we only care about really recent data, so if any messages arrive after 30 seconds, then they may be dropped. If this was a critical business process, then you would likely increase the time that messages will be guaranteed to be delivered. The length of time you choose is a trade-off between using more memory, a longer window where you are more likely to receive all messages, and less memory where messages might be lost if there is an infrastructure issue or other problems.

The second function, WriteStream, is what starts the actual stream processing and causes any messages written to the Apache Kafka topic to be brought into the Apache Spark instance and processed. Until we call WriteStream, we will not receive any actual messages.

That is the first part of our streaming application, where we process messages and take some action in real time. The action we have shown is quite simple, but you could run much more complicated commands, including joining to static datasets or even other streams, so joining stream-to-stream jobs is supported.

Aggregating the Data

In Listings 9-11 and 9-12, we show the second part of the application will take all the messages received over a period of time and aggregate the data so that it can be displayed in a dashboard. This is another common use case of streaming applications as it allows business users to see trends in real time rather than having to wait for an hourly or even daily batch process to run and update their dashboards and reports.

Listing 9-11. Aggregating time slices of data in real time using C#

```
var totalByProductSoldLast5Minutes = parsedDataFrame.
WithWatermark("timestamp", "30 seconds")
    .GroupBy(Window(Col("timestamp"), "5 minute"), Col("Product_ID")).
    Sum("Price_Sold")
    .WithColumnRenamed("sum(Price_Sold)", "Total_Price_Sold_Per_5_Minutes")
    .WriteStream()
    .Format("parquet")
    .Option("checkpointLocation", "/tmp/checkpointLocation")
    .OutputMode("append")
    .Option("path", "/tmp/ValueOfProductsSoldPer5Minutes")
    .Start();
```

Listing 9-12. Aggregating time slices of data in real time using F#

```fsharp
let totalValueSoldByProducts = parsedDataFrame.WithWatermark("timestamp",
"30 seconds")
                              |> fun dataFrame -> dataFrame.
                                 GroupBy(Functions.Window(Functions.
                                 Col("timestamp"), "5 minute"),
                                 Functions.Col("Product_ID")).Sum("Price_
                                 Sold")
                              |> fun dataFrame -> dataFrame.
                                 WithColumnRenamed("sum(Price_Sold)",
                                 "Total_Price_Sold_Per_5_Minutes")
                              |> fun dataFrame -> dataFrame.WriteStream()
                              |> fun stream -> stream.Format("parquet")
                              |> fun stream -> stream.
                                 Option("checkpointLocation", "/tmp/
                                 checkpointLocation")
                              |> fun stream -> stream.
                                 OutputMode("append")
                              |> fun stream -> stream.Option("path", "/
                                 tmp/ValueOfProductsSoldPer5Minutes")
                              |> fun stream -> stream.Start()
```

In this example, again, we have the WithWatermark, but we also have an aggregation on the DataFrame before we call WriteStream. The general approach is that we define what it is we want to happen to the data before calling WriteStream, and then Apache Spark will take care of running the aggregation and then writing the data out for us.

Waiting for stream data to appear is a bit of a different way to work with Apache Spark, and it can be harder to troubleshoot why it isn't working as expected, so it is often easier to get the DataFrame actions working with a static dataset and then copy the code over to your streaming application.

When we aggregate the data, we also use the Apache Spark Window function, which takes the name of the column that contains a timestamp that it can use and the length of time it should aggregate the data from. In this example, I have used "5 minutes," which means that for every 5-minute segment of time that we receive data, we will run the aggregation and save the data out to a file system.

Viewing the Output

In Listings 9-11 and 9-12, after starting the stream using WriteStream, we specify that we want to write the data as parquet and to append to any data that is already there. If we ran this and wrote some data to the SQL Server table, we should see data in the following format being written to disk:

```
+----------------------------------+----------+------------------------------+
|                            window|Product_ID|Total_Price_Sold_Per_5_Minutes|
+----------------------------------+----------+------------------------------+
|[2020-10-16 06:05:00, 2020-10-16 06:10:00]|       123|                 12002.759765625|
|[2020-10-16 06:55:00, 2020-10-16 06:00:00]|       123|                  6001.3798828125|
|[2020-10-16 06:00:00, 2020-10-16 06:05:00]|       123|                 9002.06982421875|
|[2020-10-16 06:10:00, 2020-10-16 06:15:00]|       123|                 9002.06982421875|
+----------------------------------+----------+------------------------------+
```

When we call StartStream, a thread is created in the background, and the processing moves onto that thread. So this means that the existing main function would finish, and our application would stop, so we need to ensure that our applications live for as long as the stream lives. In Listings 9-13 and 9-14, we show how to use the stream AwaitTermination method to keep the process alive until the stream stops.

Listing 9-13. Keeping the process alive until the streams terminate in C#

```
Task.WaitAll(
    Task.Run(() => operationalAlerts.AwaitTermination()),
    Task.Run(() => totalByProductSoldLast5Minutes.AwaitTermination())
);
```

Listing 9-14. Keeping the process alive until the streams terminate in F#

```
[|
 async {
     operationalAlerts.AwaitTermination()
 }
 async{
    totalValueSoldByProducts.AwaitTermination()
 }
```

```
|]
    |> Async.Parallel
    |> Async.RunSynchronously
    |> ignore
```

If we run our application, then we will see the aggregations written every 5 minutes or so and any rows which fail our test for a large enough margin displayed on the screen. To run the application, you will need to pass the name of the "spark-sql-kafka" JAR file for your Apache Spark version.

Summary

In this chapter, we have had a look at how we can use the, hopefully familiar, DataFrame API to create a streaming application using the two common modes of either processing each batch individually and taking some action or using Apache Spark to aggregate streaming data so that it can be used in reports and dashboards.

Similar to the DataFrame API, the Structured Streaming API provides a simple-to-use interface over what should be a technically hard thing to do well, and Apache Spark makes it almost seamless.

CHAPTER 10

Troubleshooting

In this chapter, we will take a look at how to monitor your applications and troubleshoot them. We will look at the log files that you can control and the SparkUI, a web interface for examining Apache Spark jobs, how the jobs ran in terms of performance, and what execution plan your Apache Spark job generated. In this chapter, we will not have any code samples, but we will look at the configuration and the SparkUI web interface.

Logging

Apache Spark uses log4j for its logging, and to control the amount of logging you see, you should look in your spark directory where you will find the "conf" folder, and inside you should see a log4j.properties file. If you do not see the log4j.properties file, then you should see log4j.properties.template, which you can copy to log4j.properties. If you ever make a mistake with the configuration, you can revert to the original version by copying the .template file over the top of the .properties file.

If we look inside the log4j.properties file, the first section controls how much logging we see and where the logs go to. In Listing 10-1, we will look at the first section of the log4j.properties file. For more information about log4j, see *https://logging.apache. org/log4j/2.x/*.

Listing 10-1. The first section of the log4j.properties file

```
# Set everything to be logged to the console
log4j.rootCategory=INFO, console
log4j.appender.console=org.apache.log4j.ConsoleAppender
log4j.appender.console.target=System.err
log4j.appender.console.layout=org.apache.log4j.PatternLayout
log4j.appender.console.layout.ConversionPattern=%d{yy/MM/dd HH:mm:ss} %p
%c{1}: %m%n
```

© Ed Elliott 2021
E. Elliott, *Introducing .NET for Apache Spark*, https://doi.org/10.1007/978-1-4842-6992-3_10

What this means is that the console is going to see quite a lot of messages as anything from the INFO level and above will be shown. We show the log4j message levels in Table 10-1.

Table 10-1. *The log4j message levels*

Log Level
OFF
FATAL
ERROR
WARN
INFO
DEBUG
TRACE
ALL

Typically, on my development machine, I will have the error level set to ERROR or WARN. Anything other than that results in either too few messages or far too many messages.

To change the properties file just to show ERROR rather than WARN, change the "rootCategory" line to ERROR, as shown in Listing 10-2.

Listing 10-2. Changing the logging level to ERROR

```
log4j.rootCategory=ERROR, console
```

It is also possible to control the logging level using code by using the LogLevel method on the SparkContext, which you get a reference to from the SparkSession.

Spark UI

Apache Spark comes with a UI for examining the jobs that are executed. The UI is useful as it allows us to dig into the execution of a job and troubleshoot performance issues. Knowing how to get to the Spark UI and how to diagnose performance issues is critical to becoming efficient with Apache Spark.

When you run an Apache Spark job, you may have noticed this line in the logging if you have your logging set to INFO or above:

```
INFO SparkUI: Bound SparkUI to 0.0.0.0, and started at http://machine.
dns.name:4040
```

What this means is that Apache Spark has started a web server and is listening on port 4040. While the Job is running, you can connect and view the details of the Job. However, once your Apache Spark job has finished, the web server closes down, and you are unable to view anything. In this case, the port is 4040, which is the default port, but if another instance of Apache Spark is already running, it will use port 4041 or the next free port, and the correct URL will be printed out.

As well as the Spark UI being started for each Job and then closed when the job finishes, we can request that the Apache Spark instance writes the event data that is needed for the Spark UI to a directory, and instead of the Spark UI being started when a job runs and closing down when a job finishes, you are to run a copy of what is called the history server which will read the event data from the folder that is written to by individual jobs.

To configure Apache Spark so that events are written to the folder, you should edit the spark config file conf/spark-defaults.conf and add the following two lines:

```
spark.eventLog.enabled true
spark.eventLog.dir /tmp/spark-history-logs
```

Then any data written there by jobs will be picked up by the history server, which you start by running "sbin/start-history-server.sh" from the Apache Spark installation directory. If we start the history server, then it has a different default port, and so the history server usually is available at http://localhost:18080.

History Server

There are two versions of the Spark UI, the version which starts for every instance of
Apache Spark and a way that we can store the data from previous instances, and then
we can show the details in the version of the Spark UI called the history server. Using the
history server, you can see trace output from previous instances that are no longer active.

The difference we see is that when we connect to a history server rather than the
SparkUI for one specific instance, we first arrive at an overview window that lets us drill
down to the specific instance of Apache Spark we are interested in. In Figure 10-1, we see
the history server when we first connect.

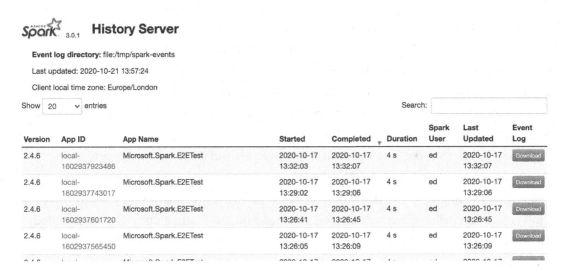

Figure 10-1. *The history server*

From the default page on the history server, we can do a couple of things. Firstly, we
can choose to download the "Event Log" which contains every log line the instance of
Apache Spark wrote but wrapped as a JSON file for simpler parsing. Secondly, we can
click an "App ID" that will take us into the main screen for the application, as shown in
Figure 10-2.

The Jobs Tab

Spark 3.0.1   Jobs   Stages   Storage   Environment   Executors   SQL   **Spark shell** application UI

Spark Jobs (?)

User: ed
Total Uptime: 29 min
Scheduling Mode: FIFO
Active Jobs: 1
Completed Jobs: 1

▸ Event Timeline

▸ **Active Jobs (1)**

▸ **Completed Jobs (1)**

Figure 10-2. *The SparkUI main screen*

The main screen is broken into several tabs across the top, and "Jobs" is the default tab. If we expand "Active Jobs," we can see if there are any currently executing jobs and what their status is. In this case, we can see that there is one active Job and one completed Job. In Figure 10-3, we can see what the active and completed jobs look like when they are expanded.

Figure 10-3. *Active and completed Jobs available to view in the SparkUI*

When we look at the description, we need to remember that the processing in Apache Spark is based on actions and transformations. Until an action occurs, the transformations are added to the plan and do not cause any actual processing. When an action occurs, in this case, the completed action was reading from a parquet file, and the running action is calling Show on a DataFrame.

Also, although it may look like two different jobs were submitted, what we are looking at is how Apache Spark converts the requests into Jobs and Stages, so this was generated by running

```
spark.Read().Parquet("/tmp/partitions").GroupBy("id").Count().Cache().
Show()
```

The last thing that the "Jobs" tab shows is a timeline of when executors were added or removed and when Jobs started, finished, and failed. We see an example of this in Figure 10-4 with two small jobs running and completing successfully around 14:25. If the jobs had failed, then the boxes would be in red.

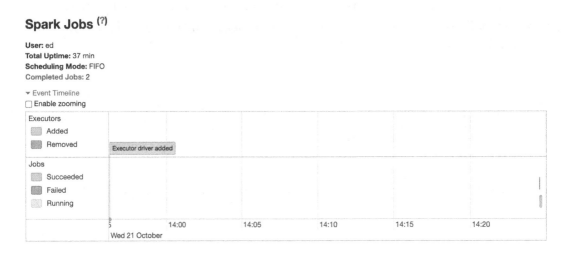

***Figure 10-4.** Spark Job timeline*

To drill down into a specific job, we can click it on the timeline view, as shown in Figure 10-4, or click the link in the description column shown in Figure 10-3. The job details tab is shown in Figure 10-5. The first thing to notice is that again we see the stages that made up the Job. The duration and the shuffle columns are used for performance troubleshooting. If one stage is taking a significant amount of time, then this is the first place we will begin to understand the performance characteristics.

Figure 10-5. *The different stages of a job broken into stages, useful for understanding which parts of the Job are slow*

When Apache Spark processes a job using the DataFrame API, Apache Spark will create a SQL execution plan and execute that plan. When we use the DataFrame API, it is an abstraction layer over the RDD API, so understanding what that plan is and how it works is a core skill when troubleshooting Apache Spark performance. In Figure 10-5, there is a link to the SQL plan that was generated for the Job. At the top, "Associated SQL Query: 0", the 0 refers to the query number, and if you click that link, it will take you to the SQL plan. We will cover the SQL plan tab and how to read the plans later on in this chapter.

In Figure 10-5, we can see another section, which we can expand, called "DAG Visualization"; the DAG is the list of operations that will be run against the RDDs. The DAG Visualization is the detail of how the RDDs are processed. Figure 10-6 shows the DAG Visualization of the Job, which includes all the operations from all of the stages that make up a job.

▶ Event Timeline

▼ DAG Visualization

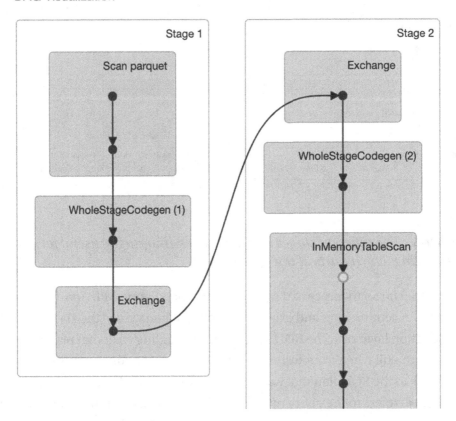

Figure 10-6. *The DAG Visualization of a job*

What the DAG in Figure 10-6 is showing us is that there were two stages in this Job. The first stage read from a parquet file, "Scan parquet", and then passed the data over to the second stage using the "Exchange" operators. Apache Spark, then, did a table scan of the data using the "InMemoryTableScan" operator. If we look at Figure 10-7, we can see that when we hover over the black and green dots, we can get even more information about what happened. In Figure 10-8, we can also look at the details of the "Scan parquet" and the "InMemoryTableScan".

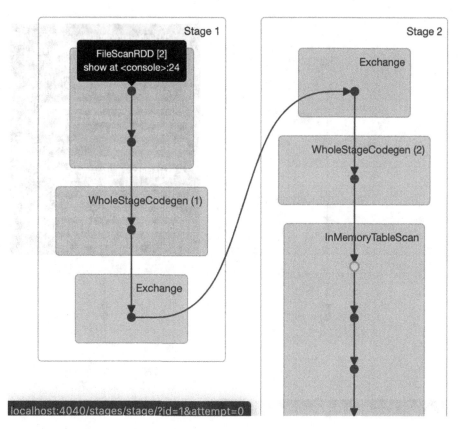

Figure 10-7. *The details of the "Scan parquet"*

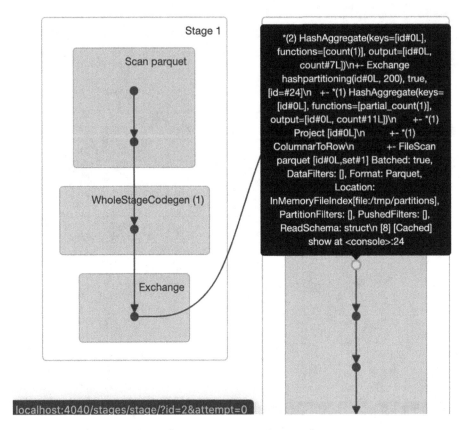

Figure 10-8. *The details of the "InMemoryTableScan"*

The job detail tab is to show how the Job is broken down into stages. After the DAG Visualization, we see the detail of how long each stage took; in Figure 10-9, we see the details of each stage, including how long each stage physically took and how many tasks made up the stage and how much data was involved.

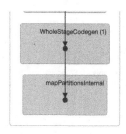

▾ Completed Stages (2)

Stage Id ▾	Description	Submitted	Duration	Tasks: Succeeded/Total	Input	Output	Shuffle Read	Shuffle Write
2	show at <console>:24 +details	2020/10/21 14:24:41	0.2 s	1/1			273.6 KiB	
1	show at <console>:24 +details	2020/10/21 14:24:33	8 s	12/12	44.1 MiB			53.9 MiB

Figure 10-9. *The details of each stage*

In Figure 10-9, we can see that there were two stages, and the table shows that the first stage was made up of 12 tasks, and the stage took 8 seconds. The second stage was made up of a single task and took 1 second. If this Job were too slow, we would use this information to start being able to narrow down exactly which task or tasks were taking the most time.

In this case, we start to see a clue as to why the Job took 8 seconds in that the first stage read in 44.1 MiB and had to do a shuffle write of 53.9 MiB. What this likely alludes to is that the way the data is being stored is not efficient for processing and has to be reordered and copied between the executors before the processing can complete.

In Figure 10-10, we click the first stage that took 8 seconds to run, which takes us to the "Stages" screen and the first stage's detail.

The Stages Tab

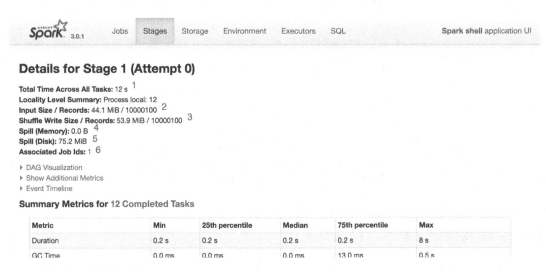

Figure 10-10. *The "Stages" tab and the stage details*

The detail we start to see in the stage details shows us

1. How long the stage took, including all of the tasks

2. How much data was read in terms of size and rows

3. How much data was "Shuffled" which is something we should try to avoid

4. How much data was spilled to memory

5. How much data was spilled to disk, which is something else we should avoid if possible

6. The Job that the stage is associated with which lets us navigate back to the job detail screen

In Figure 10-11, we move further down the stage details screen and can see a summary of the completed tasks and specifically the distribution of how long each task took. The important thing here is that we, typically, want the tasks in a stage to take about the same amount of time, so if the difference between the min and max time is wildly different, or if the difference between the 75th percentile and the max percentile is significantly different, then we likely have an issue with the way the data is partitioned. The reason is that the tasks all read a portion of the data, and if one task ends up with significantly more data than the other tasks, then the performance is bottlenecked by that one task.

▸ DAG Visualization
▸ Show Additional Metrics
▸ Event Timeline

Summary Metrics for 12 Completed Tasks

Metric	Min	25th percentile	Median	75th percentile	Max
Duration	0.2 s	0.2 s	0.2 s	0.2 s	8 s
GC Time	0.0 ms	0.0 ms	0.0 ms	13.0 ms	0.5 s
Spill (memory)	0.0 B	0.0 B	0.0 B	0.0 B	0.0 B
Spill (disk)	0.0 B	0.0 B	0.0 B	0.0 B	75.2 MiB
Input Size / Records	3.3 KiB / 0	14.1 KiB / 0	14.1 KiB / 0	14.1 KiB / 0	43.9 MiB / 10000000
Shuffle Write Size / Records	0.0 B / 0	0.0 B / 0	0.0 B / 0	0.0 B / 0	53.9 MiB / 10000000

Showing 1 to 6 of 6 entries

▾ Aggregated Metrics by Executor

Show 20 ∨ entries   Search:

Chuffle

Figure 10-11. The summary metrics for a stage broken into tasks

Here, we can see that the majority of tasks took up to 0.2 seconds, but at least one task took 8 seconds, which is quite a big difference. In Figure 10-12, we will move further down the stage details screen and see the summary metrics broken down by executors. In this example, I ran the Job on my laptop, so I have a single executor. Still, if it were on an Apache Spark cluster, then there would be likely many executors, and this would allow you to see if a specific executor caused the issue or if all of the executors had an issue. Typically, when we run Apache Spark clusters, we use a cluster made of the same type of machines. Still, nothing is stopping you from running a cluster of various machine sizes, so it might be that one executor on one node doesn't have enough memory, and that is something you can monitor on the stage details screen.

Input Size / Records	3.3 KiB / 0	14.1 KiB / 0	14.1 KiB / 0	14.1 KiB / 0	43.9 MiB / 10000000
Shuffle Write Size / Records	0.0 B / 0	0.0 B / 0	0.0 B / 0	0.0 B / 0	53.9 MiB / 10000000

Showing 1 to 6 of 6 entries

▾ Aggregated Metrics by Executor

Show 20 ∨ entries   Search:

Executor ID ▲	Logs	Address	Task Time	Total Tasks	Failed Tasks	Killed Tasks	Succeeded Tasks	Blacklisted	Input Size / Records	Shuffle Write Size / Records	Spill (Memory)	Spill (Disk)
driver		localhost:51345	13 s	12	0	0	12	false	44.1 MiB / 10000100	53.9 MiB / 10000100	0.0 B	75.2 MiB

Showing 1 to 1 of 1 entries   Previous 1 Next

Tasks (12)

Show 20 ∨ entries   Search:

Shuffle

Figure 10-12. The executor metrics

The stage details screen also gives us a visual overview of how long each task took, and as we see in Figure 10-13, if we hover over the bar, it will show the details of the specific task.

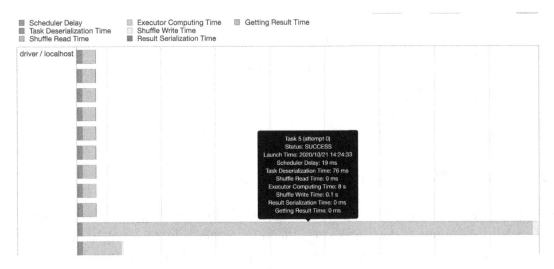

Figure 10-13. *Visualization of the performance of each task*

The additional information you see when you hover over the bar includes the task index, in this case, "Task 5," which allows us to see more details by scrolling down and examining the list of tasks shown in Figure 10-14.

Index ▲	Task ID	Attempt	Status	Locality level	Executor ID	Host	Logs	Launch Time	Duration	GC Time	Input Size / Records	Write Time	Write Size / Records
0	1	0	SUCCESS	PROCESS_LOCAL	driver	localhost		2020-10-21 14:24:33	0.2 s	13.0 ms	14.1 KiB / 0		
1	2	0	SUCCESS	PROCESS_LOCAL	driver	localhost		2020-10-21 14:24:33	0.2 s		14.1 KiB / 0		
2	3	0	SUCCESS	PROCESS_LOCAL	driver	localhost		2020-10-21 14:24:33	0.2 s		14.1 KiB / 0		
3	4	0	SUCCESS	PROCESS_LOCAL	driver	localhost		2020-10-21 14:24:33	0.2 s		14.1 KiB / 0		
4	5	0	SUCCESS	PROCESS_LOCAL	driver	localhost		2020-10-21 14:24:33	0.2 s		14.1 KiB / 0		
5	6	0	SUCCESS	PROCESS_LOCAL	driver	localhost		2020-10-21	8 s	0.5 s	43.9 MiB / 10000000	0.1 s	53.9 MiB / 10000000

Figure 10-14. *The list of tasks that made up a stage*

In Figure 10-14, we can see that the Task at Index 5, is actually "Task ID" 6, which is something to beware of.

When we look at this list of tasks, we can see the difference between the tasks that were fast and the task that was slower, that is, that the amount of data "shuffled" is much higher.

The SQL Tab

If we move now to the "SQL" tab, we can see a list of the execution plans that were generated for every query. In Figure 10-15, we can see the "SQL" tab and list of generated plans and the Jobs which the plans are associated with. In this case, there is only one plan.

Figure 10-15. *The list of tasks that made up a stage*

If we click the description, we drill down to the details of the SQL. In Figure 10-16, we can see the top of the SQL details screen, which includes how long the Job took to run and the associated Job, so you can go back and forth between SQL plan and Job. It is important to remember that when we use the DataFrame API or use SQL queries, we are effectively doing the same thing. Apache Spark parses any SQL queries and builds a plan for executing the query in the same way as DataFrame API calls cause a plan to be generated and executed. The plan generated can be the same between writing SQL queries and using the DataFrame API. Knowing that whichever way you want to approach writing code for Apache Spark causes the same plan and processing is another compelling reason to use Apache Spark.

Details for Query 0

Submitted Time: 2020/10/21 14:24:32
Duration: 9 s
Succeeded Jobs: 1

☐ Show the Stage ID and Task ID that corresponds to the max metric

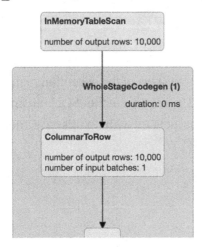

Figure 10-16. *The SQL details*

After the query details, we can see a visual representation of the query, including how many rows were output at each stage. This is useful when looking at complicated queries, particularly with joins, to help track down where rows have come from or gone missing from.

Further down the screen, in Figure 10-17, we see a textual representation of the logical plan as it is parsed and the analyzed logical plan. If you have worked with SQL Server Text plans, then these should be reasonably familiar to you. Each plan is a tree of operations with the first operation at the bottom of the tree, passing data up the tree until we get to the top of the tree.

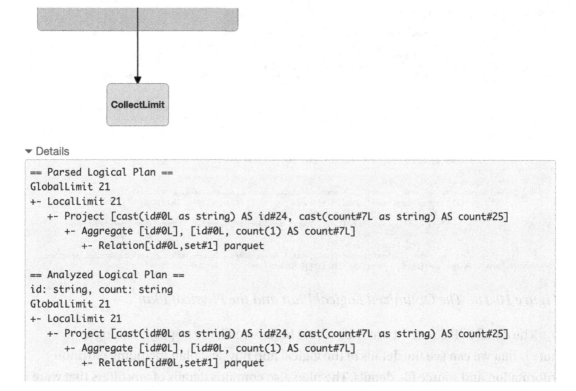

Figure 10-17. *The Parsed and Analyzed Logical Plan*

After the logical plans, in Figure 10-18, we can see the "Optimized Logical Plan," which contains the performance characteristics of the plan such as which type of join to use at each stage, and then finally the "Physical Plan," which details things like paths and partition information for how Apache Spark has to execute the plan.

```
GlobalLimit 21
+- LocalLimit 21
   +- Project [cast(id#0L as string) AS id#24, cast(count#7L as string) AS count#25]
      +- InMemoryRelation [id#0L, count#7L], StorageLevel(disk, memory, deserialized, 1 replicas)
         +- *(2) HashAggregate(keys=[id#0L], functions=[count(1)], output=[id#0L, count#7L])
            +- Exchange hashpartitioning(id#0L, 200), true, [id=#24]
               +- *(1) HashAggregate(keys=[id#0L], functions=[partial_count(1)], output=[id#0L, count#11L])
                  +- *(1) Project [id#0L]
                     +- *(1) ColumnarToRow
                        +- FileScan parquet [id#0L,set#1] Batched: true, DataFilters: [], Format: Parquet, Location:
InMemoryFileIndex[file:/tmp/partitions], PartitionFilters: [], PushedFilters: [], ReadSchema: struct<id:bigint>

== Physical Plan ==
CollectLimit 21
+- *(1) Project [cast(id#0L as string) AS id#24, cast(count#7L as string) AS count#25]
   +- *(1) ColumnarToRow
      +- InMemoryTableScan [count#7L, id#0L]
         +- InMemoryRelation [id#0L, count#7L], StorageLevel(disk, memory, deserialized, 1 replicas)
            +- *(2) HashAggregate(keys=[id#0L], functions=[count(1)], output=[id#0L, count#7L])
               +- Exchange hashpartitioning(id#0L, 200), true, [id=#24]
                  +- *(1) HashAggregate(keys=[id#0L], functions=[partial_count(1)], output=[id#0L, count#11L])
                     +- *(1) Project [id#0L]
                        +- *(1) ColumnarToRow
                           +- FileScan parquet [id#0L,set#1] Batched: true, DataFilters: [], Format: Parquet, Location:
InMemoryFileIndex[file:/tmp/partitions], PartitionFilters: [], PushedFilters: [], ReadSchema: struct<id:bigint>
```

Figure 10-18. *The Optimized Logical Plan and the Physical Plan*

The detail in Figure 10-18 may be a little hard to read; the important thing to note is that we can see the details of the logical and physical plan including partition information and source file details. The plan also contains details of any filters that were pushed down to the source.

Remaining Tabs

The remaining tabs on the Spark UI include the "Storage" tab, which is shown in Figure 10-19. The storage tab shows the details of any RDDs which have been cached. In this example, the code I wrote, which used the DataFrame API, call `DataFrame.Cache()`, which caused the data to be written to disk so that it could be used in another query without having to go through the processing again.

Figure 10-19. *The "Storage" tab overview screen*

We can see how much space the RDD takes in memory and how many of the partitions have been cached. If we click the RDD Name, it takes us to the detail of the RDD, as we can see in Figure 10-20.

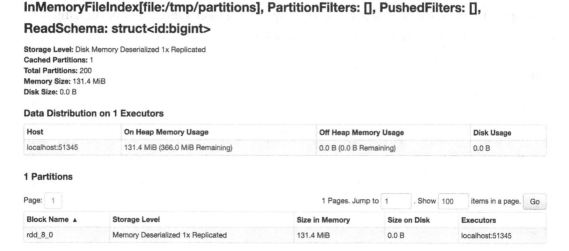

Figure 10-20. *The RDD cache detail screen*

The next tab in the Spark UI is the "Environment" tab, which includes the environment's details, such as the version of Java and Scala that the Apache Spark instance is using. The information, while useful, is hopefully not needed very often. Figure 10-21 shows the environment tab on my local Apache Spark instance.

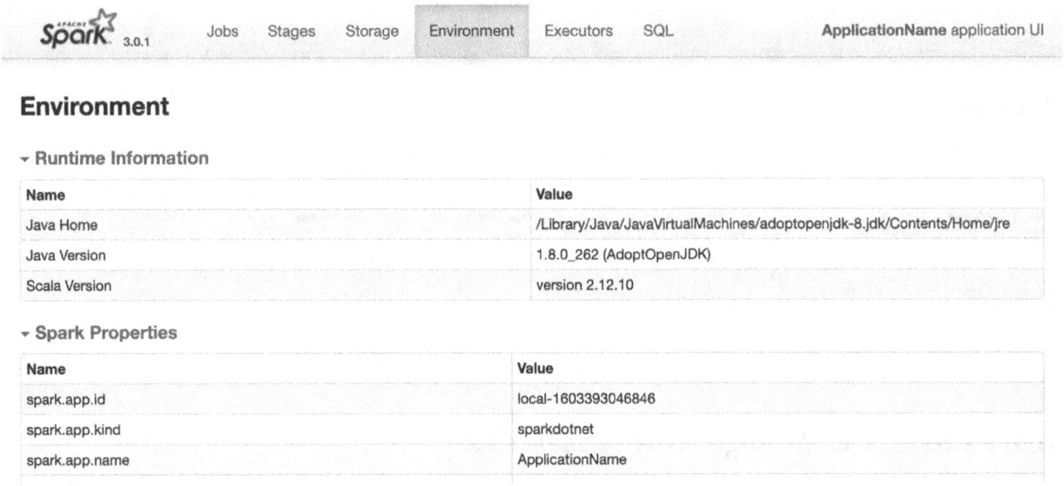

Spark 3.0.1   Jobs   Stages   Storage   Environment   Executors   SQL   **ApplicationName** application UI

Environment

▾ Runtime Information

Name	Value
Java Home	/Library/Java/JavaVirtualMachines/adoptopenjdk-8.jdk/Contents/Home/jre
Java Version	1.8.0_262 (AdoptOpenJDK)
Scala Version	version 2.12.10

▾ Spark Properties

Name	Value
spark.app.id	local-1603393046846
spark.app.kind	sparkdotnet
spark.app.name	ApplicationName

Figure 10-21. *The "Environment" tab in the Spark UI*

In Figure 10-22, we see the final tab in the Spark UI is the "Executors" tab, which shows the details of how each of the executors is performing. For example, if we see high "GC Time," then we should consider increasing the available memory or optimizing the code, so less memory is required to process the jobs.

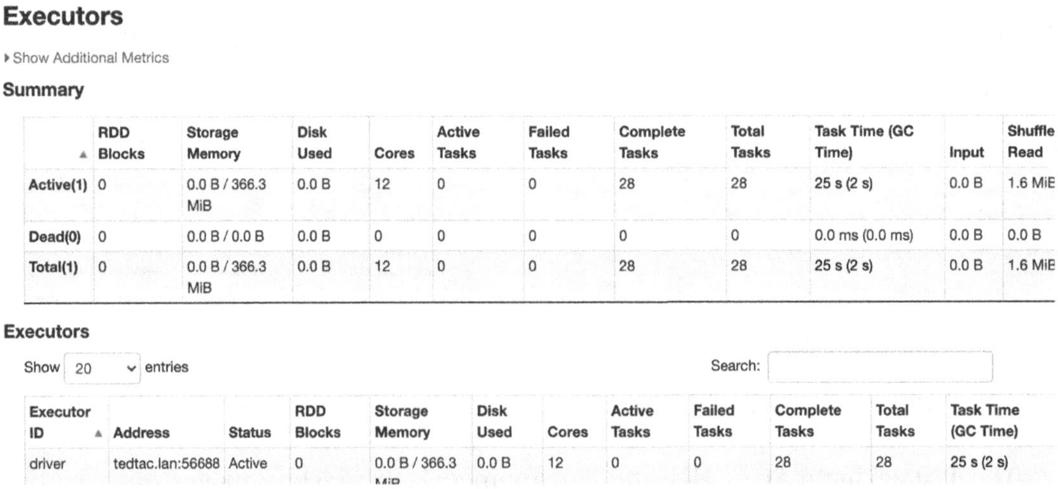

Executors

▸ Show Additional Metrics

Summary

	RDD Blocks	Storage Memory	Disk Used	Cores	Active Tasks	Failed Tasks	Complete Tasks	Total Tasks	Task Time (GC Time)	Input	Shuffle Read
Active(1)	0	0.0 B / 366.3 MiB	0.0 B	12	0	0	28	28	25 s (2 s)	0.0 B	1.6 MiE
Dead(0)	0	0.0 B / 0.0 B	0.0 B	0	0	0	0	0	0.0 ms (0.0 ms)	0.0 B	0.0 B
Total(1)	0	0.0 B / 366.3 MiB	0.0 B	12	0	0	28	28	25 s (2 s)	0.0 B	1.6 MiE

Executors

Show 20 ∨ entries   Search: [   ]

Executor ID	Address	Status	RDD Blocks	Storage Memory	Disk Used	Cores	Active Tasks	Failed Tasks	Complete Tasks	Total Tasks	Task Time (GC Time)
driver	tedtac.lan:56688	Active	0	0.0 B / 366.3 MiB	0.0 B	12	0	0	28	28	25 s (2 s)

Figure 10-22. *The "Executors" tab in the Spark UI*

The final thing to point out about the Spark UI is that when you are on a shared cluster, it can be hard to differentiate one job from another, so to help track down specific jobs, when you create your `SparkSession`, you have the option to set an identifier for the Job which allows you to quickly see the Job in the SparkUI. In Listing 10-3, we can see how to set the Application Name, and in Figure 10-23, we can see that the name is displayed in the Spark UI to help track down specific jobs.

Listing 10-3. Setting the AppName, which is then displayed in the Spark UI

```
var spark = SparkSession
    .Builder()
    .AppName("TF-IDF Application")
    .GetOrCreate();
```

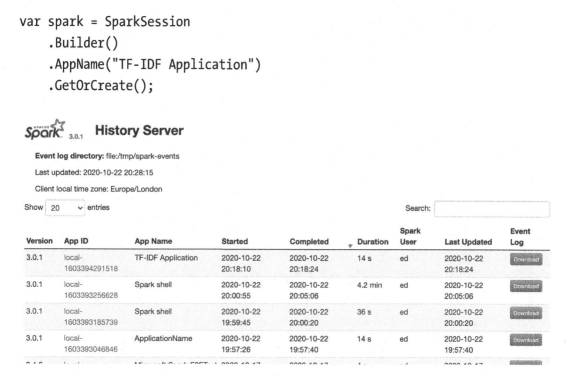

Figure 10-23. *The App Name helps differentiate Jobs when running on a shared Apache Spark instance*

Summary

In this chapter, we have had a look at the logging that you get with Apache Spark and how to configure the number of messages you see. Typically, you want to have a few messages as possible until you need to go back and troubleshoot something that failed where you need all the messages, so knowing where to configure logging is crucial.

We then had an overview of the Spark UI and the different screens we can use to get to the information we need to diagnose performance issues with our Apache Spark jobs. Hopefully, you are able to look at the Spark UI and drill down to issues without being lost in a sea of information about other, irrelevant jobs.

CHAPTER 11

Delta Lake

Delta Lake is an extension to Apache Spark created by the company behind Apache Spark, Databricks, and released as a separate open source project. Delta Lake aims to make writing to data lakes efficient in an enterprise environment – no matter which type of data lake you have, whether it is Azure Data Lake Storage, AWS S3, or Hadoop. Delta Lake brings the ACID properties that you would expect with a relational database such as Microsoft SQL Server or Oracle to a remote file system such as a data lake.

ACID

When we work with an RDBMS such as Microsoft SQL Server or Oracle, we talk about the ACID properties and how we are able to send read and write requests to the database and the RDBMS handles the ACID properties for us. The ACID properties can be described as

- **Atomicity** – Reads and writes happen in their transaction and either complete entirely or fail altogether.

- **Consistency** – Data is never left in a broken state and must be valid. The data must also pass any data constraints to ensure that not only is the data valid from a format point of view but that it meets expectations such as whether NULLs are allowed or not.

- **Isolation** – One transaction cannot affect another in-progress transaction.

- **Durability** – Once a transaction has been committed, then even if a system dies, it stays committed.

To understand what this means for Apache Spark, let us take a look at what came before, what would happen when you wrote some data out to a file system, in this case, Azure Data Lake Storage.

207

© Ed Elliott 2021
E. Elliott, *Introducing .NET for Apache Spark*, https://doi.org/10.1007/978-1-4842-6992-3_11

Consider that you have an Apache Spark application that is going to write some data using the parquet format. In the options for writing the data, you specify "overwrite," which will cause any data to be overwritten. In this scenario, you need to make sure that you are the only process writing the data as any other process will overwrite your data. Unless you have some external process to cause one and only one job to write to a specific folder, you will find you have complex timing issues you need to resolve.

Another potential issue is what happens if Apache Spark is writing to a directory and halfway through the write fails, and so it is incomplete. What should readers do? Will they know that the data is incomplete? Even if they realize that the data is incomplete, it isn't possible to go back to the previous, overwritten data.

One other issue with Apache Spark is reading from a directory that has a significant number of files. To evaluate the list of files to read in is quite expensive and can slow down jobs that read from data lakes with this problem.

With a table in a traditional RDBMS, we can read, insert, update, delete, or merge, which is a combination of insert, update, and delete. With a data lake, we have files where we can sometimes append by adding new files, or we can overwrite, but we can't open a parquet file, find some rows, and update or delete them. The traditional ways we had of working with tables in an RDBMS do not translate to a file-based data lake.

Delta Lake was introduced to fix these issues. Delta Lake brings the ACID properties from an RDBMS to a file-based data lake as well as the ability to run insert, update, delete, and even merge statements against files in a data lake, while also fixing the slow read performance from lots of files in the same directory as well as fixing the issue of multiple writers and what happens when a writer fails and leaves incomplete data. Delta Lake improves all this as well as giving you the ability to roll back to a previous version of the data.

This type of concurrency control where you keep the original data intact but provide some other way to store the versioning information is called "Multi-Version Concurrency Control" or MVCC.

The Delta Log

Delta Lake works because it creates what is called "The Delta Log," which is a set of JSON files that allow Apache Spark to read not only the data in a data lake but also the version information about which files relate to which version of the data. The delta log is in a folder called _delta_log and is made up of JSON files with a known naming system and uses locks specific to each type of data lake to ensure that the JSON files are written once and so allow multiple writers to write their data and then write to the JSON files in order, so one transaction cannot make it look like another transaction had succeeded when it actually failed or was partially overwritten.

The JSON delta log files describe the state of the data. In Listing 11-1, we can see the first JSON file written to a data lake when the folder was converted to the delta format. The listing shows the initial operation, which was "CONVERT", and then that 12 files were added to the table. The delta log also includes the schema of the written data and the paths to the file that makes up the first version.

Listing 11-1. The JSON delta log file

```json
{
    "commitInfo": {
        "timestamp": 1603400609668,
        "operation": "CONVERT",
        "operationParameters": {
            "numFiles": 12,
            "partitionedBy": "[]",
            "collectStats": false
        },
        "operationMetrics": {
            "numConvertedFiles": "12"
        }
    }
}
{
    "protocol": {
        "minReaderVersion": 1,
        "minWriterVersion": 2
    }
}
{
    "metaData": {
        "id": "8942a94c-506b-488c-8247-0da4e861a37a",
        "format": {
            "provider": "parquet",
            "options": {}
        },
```

```
            "schemaString": "{\"type\":\"struct\",\"fields\":[{\"name\":\"id\",
            \"type\":\"long\",\"nullable\":true,\"metadata\":{}}]}",
            "partitionColumns": [],
            "configuration": {},
            "createdTime": 1603400609648
        }
    }
    {
        "add": {
            "path": "part-00011-707f035c-4ddb-461f-9d52-bc1f41f1f08c-c000.
            snappy.parquet",
            "partitionValues": {},
            "size": 804,
            "modificationTime": 1603400607000,
            "dataChange": true
        }
    }
    {
        "add": {
            "path": "part-00000-707f035c-4ddb-461f-9d52-bc1f41f1f08c-c000.
            snappy.parquet",
            "partitionValues": {},
            "size": 796,
            "modificationTime": 1603400607000,
            "dataChange": true
        }
    }
```

Reading Data

To read the data from this delta format, Apache Spark goes to the "_delta_log" folder and reads each of the JSON files in name order and then evaluates which data files should be included in the DataFrame that is returned. Having a set of JSON files tell Apache Spark which files to read means that Apache Spark does not have to enumerate all the files, with the performance overhead that it attracts.

The order that Apache Spark writes the data is essential. First of all, the actual parquet files are written, and then lastly the JSON file is updated. This means if Apache Spark writes the data files and then crashes, the delta format is not left in an inconsistent state; the parquet files without a reference in the delta log will be ignored.

Changing Data

In this section, we will look at how Delta Lake modifies the data that is available to read. We will first give an overview of all the different features of Delta Lake and then finish up with an example Delta Lake application that shows off the full range of additional features. We will walk through a demo of using Delta Lake in C# and then in F#.

Appending Data

The most straightforward operation is to add more data to an existing dataset, and Apache Spark does that by writing the new data files and then adding to the JSON more of the "add" directives that we saw in Listing 11-1. In Listing 11-2, we can see where after running an "Append" we get the next JSON file with the next set of files' details to add to the DataFrame.

Listing 11-2. Appending more data causes additional "add" directives to be added via a JSON delta log file

```
{
    "commitInfo": {
        "timestamp": 1603740783553,
        "operation": "WRITE",
        "operationParameters": {
            "mode": "Append",
            "partitionBy": "[]"
        },
        "readVersion": 0,
        "isBlindAppend": true,
        "operationMetrics": {
            "numFiles": "12",
            "numOutputBytes": "5780",
            "numOutputRows": "50"
        }
    }
}
```

```
{
    "add": {
        "path": "part-00000-29d1078d-45f1-40f2-8058-1dc16eee7bf2-c000.
        snappy.parquet",
        "partitionValues": {},
        "size": 481,
        "modificationTime": 1603740783000,
        "dataChange": true
    }
}
{
    "add": {
        "path": "part-00001-1b7ad20f-a037-4ca4-a786-be6400e5b3b1-c000.
        snappy.parquet",
        "partitionValues": {},
        "size": 481,
        "modificationTime": 1603740783000,
        "dataChange": true
    }
}
```

If we asked Apache Spark to read now, then what it would do is evaluate the first JSON file, find all the "add" directives, then evaluate the second JSON file and evaluate the "add" directives in the second file, and create a DataFrame from all of the underlying parquet files.

Overwriting Data

When we want to overwrite the data so that the data we are writing becomes the complete dataset, Apache Spark marks the write as an overwrite, and all previous parquet files are ignored.

In Listing 11-3, we can see that the "Overwrite" operation caused the parquet files to be removed from the JSON delta log using the "remove" directive and new files added using the "add" directive. This means that the data can be read from the underlying parquet files from any point in time, which does not change.

Listing 11-3. The "remove" directives are causing Apache Spark to ignore the underlying files

```
{
    "commitInfo": {
        "timestamp": 1603741047813,
        "operation": "WRITE",
        "operationParameters": {
            "mode": "Overwrite",
            "partitionBy": "[]"
        },
        "readVersion": 1,
        "isBlindAppend": false,
        "operationMetrics": {
            "numFiles": "12",
            "numOutputBytes": "5780",
            "numOutputRows": "50"
        }
    }
}
{
    "remove": {
        "path": "part-00005-6bd2f7c1-a364-4028-9846-3da01fd36f7f-c000.
        snappy.parquet",
        "deletionTimestamp": 1603741047812,
        "dataChange": true
    }
}
{
    "remove": {
        "path": "part-00001-437c2c31-ff49-489b-aff8-274f3b3de4b2-c000.
        snappy.parquet",
        "deletionTimestamp": 1603741047813,
        "dataChange": true
    }
}
```

```
{
    "add": {
        "path": "part-00000-061e39ea-a20c-4671-8abb-adae938d2115-c000.
        snappy.parquet",
        "partitionValues": {},
        "size": 481,
        "modificationTime": 1603741047000,
        "dataChange": true
    }
}
{
    "add": {
        "path": "part-00001-8428416f-4c93-4d5d-bbcb-830905e1221f-c000.
        snappy.parquet",
        "partitionValues": {},
        "size": 481,
        "modificationTime": 1603741047000,
        "dataChange": true
    }
}
```

Changing Data

Until now, we have looked at how to append additional files or overwrite the entire Delta table, which is relatively trivial. What if we want to edit the data within the Delta table? How does Delta Lake work with that?

If we want to delete a row or update a row, then it is much more complicated. Delta Lake will read the table's current state and then read in the data to identify rows to either change or delete. Once Delta Lake knows which specific rows need to be deleted or changed, it will create a new file containing the rows that were meant to be left as they were and any updated rows. Any rows that were due to be deleted are not written to the new file. When the new file has been written, an "add" directive for the new file and a "remove" directive for the previous file are written to the delta log files.

This means that if you have a Delta table that is made up of a single parquet file with 1 million rows and you change one single row, then the other 999,999 rows are rewritten along with the newly modified file. This is a waste, but it is the only available option because parquet is not an updateable format. In reality, if you were not using the Delta Lake format, you would still incur the high cost as you would have to overwrite the files, so even though it is not optimal, it is no less optimal than the other possible solutions.

In Listing 11-4, we see the result of an update statement on the delta log.

Listing 11-4. Result of the delta log after running an update statement

```
{
    "commitInfo": {
        "timestamp": 1603742054064,
        "operation": "UPDATE",
        "operationParameters": {
            "predicate": "(id#529L > 500)"
        },
        "readVersion": 1,
        "isBlindAppend": false,
        "operationMetrics": {
            "numRemovedFiles": "6",
            "numAddedFiles": "6",
            "numUpdatedRows": "499",
            "numCopiedRows": "1"
        }
    }
}
{
    "remove": {
        "path": "part-00006-d4e299fc-0ae4-48d8-9252-b00b3b78584d-c000.
        snappy.parquet",
        "deletionTimestamp": 1603742053905,
        "dataChange": true
    }
}
```

```
{
    "remove": {
        "path": "part-00010-d4e299fc-0ae4-48d8-9252-b00b3b78584d-c000.
        snappy.parquet",
        "deletionTimestamp": 1603742053905,
        "dataChange": true
    }
}
{
    "add": {
        "path": "part-00004-426a4814-9962-4aa5-81da-e38480d86c5c-c000.
        snappy.parquet",
        "partitionValues": {},
        "size": 493,
        "modificationTime": 1603742054000,
        "dataChange": true
    }
}
{
    "add": {
        "path": "part-00005-b6e1b947-3109-4951-af4a-f01a894642ff-c000.
        snappy.parquet",
        "partitionValues": {},
        "size": 493,
        "modificationTime": 1603742054000,
        "dataChange": true
    }
}
```

The update statement ran and removed any file that contained a matching row and wrote a new file with any new data. The interesting thing is how readable the delta log is for humans. In this example, it even recorded the update operation and the filter that was used to find rows to modify.

Checkpoints

If you have many modifications to a dataset stored in the delta format, you might find that it starts to get slow to enumerate all these JSON files. Another feature of Delta Lake is that it uses "checkpoint" files to store the state of one particular version in a format that is even faster to read. When Delta Lake decides it is appropriate, it will create a parquet file with the current state, and in the "_last_checkpoint" file, it records the checkpoint's version. If you read the checkpoint file as a parquet file and show the data, then you will see something that looks like this:

```
+----+-------------------+------+-------------------+--------+----------+
| txn|                add|remove|           metaData|protocol|commitInfo|
+----+-------------------+------+-------------------+--------+----------+
|null|[part-00007-d4e29...|  null|               null|    null|      null|
|null|[part-00010-d4e29...|  null|               null|    null|      null|
|null|[part-00009-d4e29...|  null|               null|    null|      null|
|null|[part-00002-d4e29...|  null|               null|    null|      null|
|null|[part-00004-d4e29...|  null|               null|    null|      null|
|null|[part-00006-d4e29...|  null|               null|    null|      null|
|null|[part-00008-d4e29...|  null|               null|    null|      null|
|null|[part-00003-d4e29...|  null|               null|    null|      null|
|null|[part-00000-d4e29...|  null|               null|    null|      null|
|null|[part-00011-d4e29...|  null|               null|    null|      null|
|null|[part-00001-d4e29...|  null|               null|    null|      null|
|null|               null|  null|               null|  [1, 2]|      null|
|null|               null|  null|[3434f91f-fb53-4e...|    null|      null|
|null|[part-00005-d4e29...|  null|               null|    null|      null|
+----+-------------------+------+-------------------+--------+----------+
```

History

Because of the way the data is written in individual parquet files and then the delta log is used to keep a record of how the data should look at any point in time, we can also view the history of the Delta table and select data as if it was either a specific version or as if it was a specific time.

To view the available history, we could manually examine the JSON delta log, which is documented by Databricks at `https://databricks.com/blog/2019/08/21/diving-into-delta-lake-unpacking-the-transaction-log.html#:~:text=The%20Delta%20Lake%20Transaction%20Log%20at%20the%20File%20Level&text=Each%20commit%20is%20written%20out,json%20%2C%20the%20following%20as%20200000002`, or the protocol itself at `https://github.com/delta-io/delta/blob/master/PROTOCOL.md`.

Delta Lake also gives us an API we can call, either using objects or SQL requests. When we use the API, we are given a DataFrame with the details of the Delta table's history.

If we look at the contents of the DataFrame, then we will see

- The version

- The timestamp

- The user who created the version

- The operations, WRITE, DELETE, UPDATE, and so on

- The job or the notebook details

- The Cluster ID

- Details of the operation such as the filter used and the number of files added and removed

Vacuum

Because every time a change is made, more files are added, it might be that for a table that is updated often, the size of the data becomes too expensive to store forever. To cater to this, Delta Lake includes a way to set a retention history length and then run Vacuum, which will delete any files that are no longer needed to serve the history.

If you specify a period of time, such as seven days, which is the default, then any files needed to create the history from the last seven days are kept. This might mean that you have a file from over seven days ago, which is still used to provide the Delta table's current version. If you consider that a file is only obsolete if the data in it is deleted or updated, if no data in a file is changed, then the file will still be valid, even though you specify that you want to Vacuum data older than seven days.

We can issue a Vacuum command whenever we like, but we should beware that the files will stay there indefinitely if we do not run a Vacuum command.

Vacuum clears up the data files. Any log files are kept until after a checkpoint, which happens automatically after every ten commits.

Merge

With the Delta Lake format and the ability to modify existing files, logically if not physically, the Apache Spark team also introduced merge for Delta Lake, which means we can have a single process that can

- Update rows

- Delete rows

- Insert new rows

Merge is an exciting feature because we can ask Apache Spark to run multiple operations in a single pass, rather than manually running an Update, Insert, and Delete.

Schema Evolution

The Delta Lake format includes the schema of the underlying data, which means that if we try to do things like appending a new data file with an extra column, the append will fail. To fix this, you can include the "mergeSchema" option, which automatically updates the schema with the new column. For any old rows, the new column will be null. However, for any newly appended data, the column will have data.

Time Travel

The history of the table allows us to see which version exists, and then, with the DataFrame API, we can specify options to control the version, either "timestampAsOf" which pulls the data as if it was a specific time or "versionAsOf" which pulls data as if it was a specific version.

This ability to quickly go back in time is handy for troubleshooting. It has saved me personally a couple of times from having to reload large quantities of data by being able to roll back to a specific date and time and then rerun previously broken data pipelines.

Example Delta Lake Applications

Hopefully, by now, you have a good appreciation for what exactly Delta Lake is. So in this section, we will look at how to add the Delta Lake code to an Apache Spark instance and create our .NET for Apache Spark applications that utilize the Delta Lake format.

Configuration

Delta Lake is not part of the core Apache Spark project. The core team has created it, but it is viewed as if it was a third-party component. Because it does not ship with Apache Spark, there are few things we need to do.

Firstly, we need to ensure our Apache Spark instance loads the JAR file for Delta Lake. I do this by modifying the $SPARK_HOME/conf/spark-defaults.conf file and adding this line:

```
spark.jars.packages io.delta:delta-core_2.12:0.7.0
```

This causes the JAR file to be downloaded and included in every subsequent Apache Spark instance that starts.

Now Apache Spark has the Delta Lake format, and we need to include the .NET objects as they also are not part of the main Microsoft.Spark NuGet package. The Delta Lake .NET objects are in the Microsoft.Spark.Extensions.Delta NuGet package, so you will need to add that to your project.

Finally, so that we can use the Delta Lake SQL extensions, we will need to tell Apache Spark to enable the extended SQL commands using the configuration option "spark.sql. extensions" which is set to "io.delta.sql.DeltaSparkSessionExtension".

CSharp

This section will go through an example of how to use the Delta Lake extension in .NET for Apache Spark. First, we will go through the C# example and then the F# example.

In Listing 11-5, we create the SparkSession but using the config options to specify that we want to use the DeltaSparkSessionExtensions.

Listing 11-5. Create the SparkSession passing in the details of the Delta extension we wish to load

```
var spark = SparkSession.Builder()
    .Config("spark.sql.extensions", "io.delta.sql.
DeltaSparkSessionExtension")
    .GetOrCreate();
```

We then, as a one-off exercise, create the Delta Lake table. With Delta Lake, there is a static class called DeltaTable, which gives us some useful methods for getting a reference to a delta table and converting a parquet file into the Delta Lake format. In Listing 11-6, we use DeltaTable.IsDeltaTable to see if the delta table exists or if we have to write a parquet file and then convert it to the Delta Lake format.

Listing 11-6. Converting a parquet file to Delta Lake

```
if (!DeltaTable.IsDeltaTable("parquet.`/tmp/delta-demo`"))
{
    spark.Range(1000).WithColumn("name", Lit("Sammy")).Write().
Mode("overwrite").Parquet("/tmp/delta-demo");
    DeltaTable.ConvertToDelta(spark, "parquet.`/tmp/delta-demo`");
}
```

Note that in Listing 11-6, we are explicitly converting a parquet file into the Delta Lake format, but we could have written a DataFrame using the "delta" format.

In Listing 11-7, we use `Delta.ForPath` to get a reference to the Delta table and then convert the reference to the Delta table into a `DataFrame` by using `ToDF()`.

Listing 11-7. Using DeltaTable to get a reference to the DataFrame

```
var delta = DeltaTable.ForPath("/tmp/delta-demo");
delta.ToDF().OrderBy(Desc("Id")).Show();
```

Appending data to the Delta table is as simple as it is with other formats. We specify the mode as "append"; we show this in Listing 11-8.

Listing 11-8. Appending data to a Delta table

```
spark.Range(5, 500 ).WithColumn("name", Lit("Lucy")).Write().
Mode("append").Format("delta").Save("/tmp/delta-demo");
```

If we want to update the data in the Delta table, we need to use the reference to the `DeltaTable` rather than the `DataFrame` that is returned from `DeltaTable.ToDF`. In Listing 11-9, I show how to use the `DeltaTable` reference to update the Delta table. In this example, we find any row that has an id greater than 500 and then set the id column to the value 999.

Listing 11-9. Updating a Delta table

```
delta.Update(Expr("id > 500"), new Dictionary<string, Column>()
{
    {"id", Lit(999)}
});
```

Deleting from the Delta table is done by calling `DeltaTable.Delete` and then passing in a filter. If you do not give in a filter, then every row is deleted. Listing 11-10 shows the `Delete` operation.

Listing 11-10. Deleting from a Delta table

```
delta.Delete(Column("id").EqualTo(999));
```

To view the history from a table, we use the DeltaTable.History() method which returns a DataFrame, so we can call Show or do any filtering you need to do. This is useful because you can filter the DataFrame to find one specific update and then use the version and/or time details to read from the table at the specific update. Listing 11-11 shows how to request the history of a Delta table.

Listing 11-11. Requesting the history from a Delta table

```
delta.History().Show(1000, 10000);
```

Now you have the available history; you can use the "timestampAsOf" and "versionAsOf" options to specify which exact version of the Delta table you want rather than the latest version. In Listing 11-12, we show how to read the Delta table as if it was a specific version or time.

Listing 11-12. Reading from the Delta table using time travel

```
spark.Read().Format("delta").Option("versionAsOf", 0).Load("/tmp/delta-
demo").OrderBy(Desc("Id")).Show();

spark.Read().Format("delta").Option("timestampAsOf", "2021-10-22
22:03:36").Load("/tmp/delta-demo").OrderBy(Desc("Id")).Show();
```

The next operation we will look at is the merge operation. When we merge data in, we use the Delta table as the target and a DataFrame as the source. To make it easier, it is often better to alias both tables. In Listing 11-13, I show an example merge operation, and I alias the Delta table as "target" and the DataFrame as "source", so when providing the filter to show which rows to match, we can provide the filter in a simple string format "target.id = source.id". When we have supplied the filter, then we can optionally provide up to two actions when the rows match and a single action for when a row does not match. When a row matches, we can also optionally provide a second filter and either update the matching rows or delete them.

When we do not match any rows, we can either insert all the columns or provide a list of columns. In this case, because I have the same schema in the Delta table and the DataFrame, InsertAll works fine.

Listing 11-13. The Delta Lake merge operation

```
var newData = spark.Range(10).WithColumn("name", Lit("Ed"));

delta.Alias("target")
        .Merge(newData.Alias("source"), "target.id = source.id")
        .WhenMatched(newData["id"].Mod(2).EqualTo(0)).Update(new
        Dictionary<string, Column>()
                                                        {
                                                            {"name",
                                                            newData
                                                            ["name"]}
                                                        })
        .WhenMatched(newData["id"].Mod(2).EqualTo(1)).Delete()
        .WhenNotMatched().InsertAll()
    .Execute();
```

The logic behind the example in Listing 11-13 is

- Find any row in the Delta table that has the matching id in the DataFrame.

- If any rows match and they are even numbers, `Mod(2).EqualTo(0)`, update the row and set the name column to the value of the name column for the associated id in the source DataFrame.

- If any rows match and they are odd numbers, `Mod(2).EqualTo(1)`, delete the row.

- If any ids in the source DataFrame do not already exist in the target Delta table, then insert all the columns from the source DataFrame into the target Delta table.

Finally, in Listing 11-14, we show the final operation to demonstrate is the `Vacuum` method, which tidies up, if it can, any older data files that are not needed to support the current version and any version within the retention period.

Listing 11-14. Delta table Vaccum

```
delta.Vacuum(1F)
```

FSharp

This section will go through an example of how to use the Delta Lake extension in .NET for Apache Spark using F#.

In Listing 11-15, we create the `SparkSession` but using the config options to specify that we want to use the `DeltaSparkSessionExtensions`.

Listing 11-15. Create the SparkSession passing in the details of the Delta extension we wish to load

```
let spark = SparkSession.Builder()
            |> fun builder -> builder.Config("spark.sql.extensions",
               "io.delta.sql.DeltaSparkSessionExtension")
            |> fun builder -> builder.GetOrCreate()
```

We then, as a one-off exercise, create the Delta Lake table. With Delta Lake, there is a static class called DeltaTable, which gives us some useful methods for getting a reference to a delta table and converting a parquet file into the Delta Lake format. In Listing 11-16, we use DeltaTable.IsDeltaTable to see if the delta table exists or if we have to write a parquet file and then convert it to the Delta Lake format.

Listing 11-16. Converting a parquet file to Delta Lake

```
let delta = match DeltaTable.IsDeltaTable("parquet.`/tmp/delta-demo`") with
                | false -> spark.Range(1000L)
                            |> fun dataframe -> dataframe.WithColumn("name",
                                Functions.Lit("Sammy"))
                            |> fun dataframe -> dataframe.Write()
                            |> fun writer -> writer.Mode("overwrite").
                                Parquet("/tmp/delta-demo")
                            DeltaTable.ConvertToDelta(spark, "parquet.`/tmp/
                            delta-demo`")
                | _ -> DeltaTable.ForPath("/tmp/delta-demo")
```

Note that in Listing 11-16, we are explicitly converting a parquet file into the Delta Lake format, but we could have written a DataFrame using the "delta" format. If we do not have to create the Delta table, we use `Delta.ForPath` to get a reference to the Delta table and then convert the reference to the Delta table into a `DataFrame` by using `ToDF()`.

Appending data to the Delta table is as simple as it is with other formats. We specify the mode as "append"; we show this in Listing 11-17.

Listing 11-17. Appending data to a Delta table

```
spark.Range(5L, 500L)
    |> fun dataframe -> dataframe.WithColumn("name", Functions.Lit("Lucy"))
    |> fun dataframe -> dataframe.Write()
    |> fun writer -> writer.Mode("append").Format("delta").Save("/tmp/
        delta-demo")
```

If we want to update the data in the Delta table, we need to use the reference to the `DeltaTable` rather than the `DataFrame` that is returned from `DeltaTable.ToDF`. In Listing 11-18, I show how to use the `DeltaTable` reference to update the Delta table. In this example, we find any row that has an id greater than 500 and then set the id column to the value 999.

Listing 11-18. Updating a Delta table

```
delta.Update(Functions.Expr("id > 500"), Dictionary<string, Column>(dict
[("id", Functions.Lit(999))]))
```

Deleting from the Delta table is done by calling `DeltaTable.Delete` and then passing in a filter. If you do not give in a filter, then every row is deleted. Listing 11-19 shows the `Delete` operation.

Listing 11-19. Deleting from a Delta table

```
delta.Delete(Functions.Col("id").EqualTo(999))
```

To view the history from a table, we use the DeltaTable.History() method which returns a DataFrame, so we can call Show or do any filtering you need to do. This is useful because you can filter the DataFrame to find one specific update and then use the version and/or time details to read from the table at the specific update. Listing 11-20 shows how to request the history of a Delta table.

Listing 11-20. Requesting the history from a Delta table

```
delta.History()
        |> fun dataframe -> dataframe.Show()
```

Now you have the available history; you can use the "timestampAsOf" and "versionAsOf" options to specify which exact version of the Delta table you want rather than the latest version. In Listing 11-21, we show how to read the Delta table as if it was a specific version or time.

Listing 11-21. Reading from the Delta table using time travel

```
spark.Read()
    |> fun reader -> reader.Format("delta")
    |> fun reader -> reader.Option("versionAsOf", OL)
    |> fun reader -> reader.Load("/tmp/delta-demo")
    |> fun dataframe -> dataframe.OrderBy(Functions.Desc("id"))
    |> fun ordered -> ordered.Show()
```

```
spark.Read()
    |> fun reader -> reader.Format("delta")
    |> fun reader -> reader.Option("timestampAsOf", "2022-01-01")
    |> fun reader -> reader.Load("/tmp/delta-demo")
    |> fun dataframe -> dataframe.OrderBy(Functions.Desc("id"))
    |> fun ordered -> ordered.Show()
```

The next operation we will look at is the merge operation. When we merge data in, we use the Delta table as the target and a DataFrame as the source. To make it easier, it is often better to alias both tables. In Listing 11-22, I show an example merge operation, and I alias the Delta table as "target" and the DataFrame as "source", so when providing the filter to show which rows to match, we can provide the filter in a simple string format "target.id = source.id". When we have supplied the filter, then we can optionally provide up to two actions when the rows match and a single action for when a row does not match. When a row matches, we can also optionally provide a second filter and either update the matching rows or delete them.

When we do not match any rows, we can either insert all the columns or provide a list of columns. In this case, because I have the same schema in the Delta table and the DataFrame, InsertAll works fine.

Listing 11-22. The Delta Lake merge operation

```
let newData = spark.Range(10L)
                |> fun dataframe -> dataframe.WithColumn("name",
                    Functions.Lit("Ed"))
                |> fun newData -> newData.Alias("source")

delta.Alias("target")
    |> fun target -> target.Merge(newData, "source.id = target.id")
    |> fun merge -> merge.WhenMatched(newData.["id"].Mod(2).EqualTo(0))
    |> fun evens -> evens.Update(Dictionary<string, Column>(dict [("name",
        newData.["name"])]))
    |> fun merge -> merge.WhenMatched(newData.["id"].Mod(2).EqualTo(0))
    |> fun odds -> odds.Delete()
    |> fun merge -> merge.WhenNotMatched()
    |> fun inserts -> inserts.InsertAll()
    |> fun merge -> merge.Execute()
```

The logic behind the example in Listing 11-22 is

- Find any row in the Delta table that has the matching id in the DataFrame.

- If any rows match and they are even numbers, `Mod(2).EqualTo(0)`, update the row and set the name column to the value of the name column for the associated id in the source DataFrame.

- If any rows match and they are odd numbers, `Mod(2).EqualTo(1)`, delete the row.

- If any ids in the source DataFrame do not already exist in the target Delta table, then insert all the columns from the source DataFrame into the target Delta table.

Finally, in Listing 11-23, we show the final operation to demonstrate is the `Vacuum` method, which tidies up, if it can, any older data files that are not needed to support the current version and any version within the retention period.

Listing 11-23. Delta table Vaccum

```
delta.Vacuum(1F)
```

Summary

In this chapter, we have taken a look at Delta Lake and how it can help us create data applications in Data Lakes. I have used Delta Lake in production, and the benefits such as being able to roll back a table to a specific point in time and update, delete, and merge in data to existing Delta tables are really compelling.

APPENDIX A

Running in the Cloud

When we have developed our .NET for Apache Spark application, we will likely want to run it on a production cluster, which means running in the cloud for many organizations today. When we want to run our application in the cloud, we have three options:

1. Use Databricks, which is a fully managed Apache Spark instance by the creators of Apache Spark.

2. Use a third-party implementation such as AWS EMR, Azure Synapse Analytics, or GCP Dataproc.

3. Create and manage your clusters.

   In this appendix, we will take a look at what Databricks is and how we can use it to manage and execute our .NET for Apache Spark applications.

Databricks

Databricks is available in Amazon Web Services (AWS) and Azure but not available in the Google Cloud Platform (GCP). Databricks is a fully managed implementation of Apache Spark. Databricks includes various functionality such as user management, job management, an interactive UI, and performance enhancements that are not in the open source version of Apache Spark on the third-party implementations such as AWS EMR and Azure Synapse Analytics.

Both the AWS EMR and the Azure Synapse Analytics are platforms for running "big data" processing jobs and can run Apache Spark jobs. Neither of the third-party platforms is exclusively Apache Spark, so they are quite a compelling use case for people who need more than Apache Spark in a single instance.

© Ed Elliott 2021
E. Elliott, *Introducing .NET for Apache Spark*, https://doi.org/10.1007/978-1-4842-6992-3

Running your Virtual Machines and building your clusters to run Apache Spark jobs is the least likely option to be chosen today, and this is out of scope for this appendix. However, this URL from Apache Spark gives a good overview of creating and managing your cluster for running Apache Spark jobs: *https://spark.apache.org/docs/latest/cluster-overview.html*.

Approaches to Executing Code on Databricks

When we use Databricks, we have a choice as to whether we use notebooks which are interactive web UI that includes code and the output from running code. Notebooks can be shared and deployed to a Databricks instance. Notebooks can be run manually or as part of an automated Job. Notebooks on Databricks currently support Scala, Python, and R but unfortunately not .NET; however, we can deploy our .NET for Apache Spark applications and then call those from notebooks, so it is possible to at least run the code via a notebook.

Aside from using notebooks, we can deploy applications and run them from Jobs in Databricks or run them from external tools such as Azure Data Factory or Apache Airflow, which are both ETL orchestration tools.

Creating a Databricks Workspace

This appendix will not discuss how to create a Databricks workspace in detail. In brief, the process is different depending on whether you want to build your Databricks workspace on AWS or Amazon. To build your workspace on AWS, you go to *https://databricks.com/*, sign up for an account, and connect that account to your AWS account. If you would like to build your workspace on Azure, you visit the Azure marketplace and create a new "Azure Databricks" instance. Finally, it is possible to create a cut-down version of Databricks using their "Community" offering, which gives you access to a single cluster for a limited amount of time each month with less functionality than the full Databricks workspaces.

In Figure A-1, we can see the Databricks UI, which uses the menu on the left-hand side to navigate to the different sections, which are

1. The home screen

2. The workspace explorer

3. Recent notebooks

4. Hive databases and tables

5. Cluster management

6. Job management

7. ML model management

8. Search page

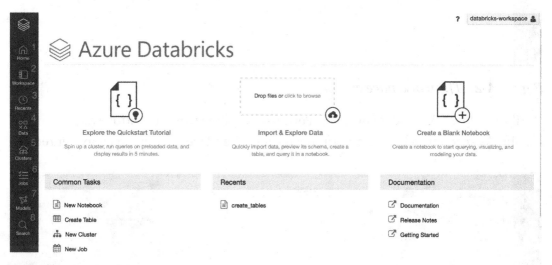

Figure A-1. *The main Databricks UI*

Home Screen

The home screen shows a list of recently used notebooks and also shortcuts to create notebooks, hive tables, clusters, and jobs.

Workspace Explorer

The Databricks workspace is like a file system, containing a list of folders which store notebooks. In Figure A-2, we can see the workspace screen and a notebook that has been created. Every user in a workspace has their own "home" folder, but you can also create shared folders. It is possible to manage the security on a folder to limit which users have access.

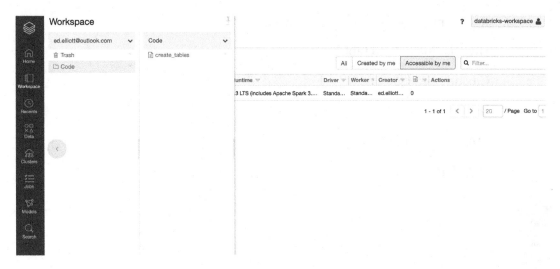

Figure A-2. *The workspace explorer*

If we right-click a folder and click "Permissions," then we can control which Databricks accounts have access to a specific folder, and we see the screen in Figure A-3.

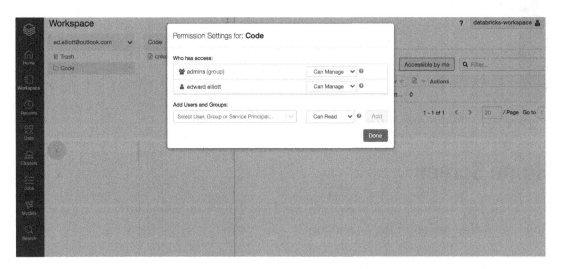

Figure A-3. *Setting the permissions on a folder in a Databricks workspace*

Data

The data screen shows a list of Hive databases and tables; if we drill down to a table as we can see in Figure A-4, we can see the schema for a table and a sample of the data. If the table was a Delta table, that is, it was created using the delta format, we would additionally see the history of the table data.

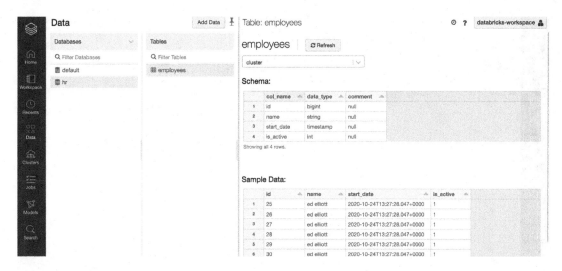

Figure A-4. *The data screen*

Additionally, in the data screen, we can create new tables by uploading data or pointing to some existing data and change the permissions on databases as well as deleting tables and databases.

Clusters

The cluster management screen allows us to create, manage, and destroy clusters. Typically, with a Databricks workspace, we will use a cluster for running Notebooks, which can be a shared cluster, and when we want to run a production job, either start up a cluster for our applications, run the process, and then destroy the cluster or keep a separate cluster running for any production jobs. Each cluster can self-terminate or shut down after a period of inactivity, or if you constantly have processing to run, such as when you are running streaming applications, you can keep a cluster running.

In Figure A-5, we can see the cluster management screen, and in this case there is a single cluster and it is running.

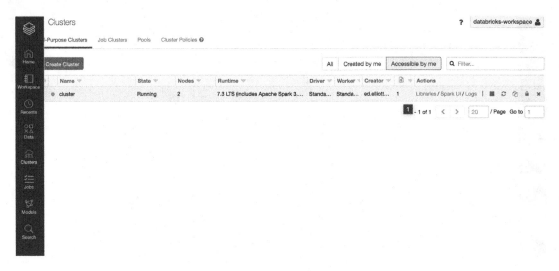

Figure A-5. *The cluster management screen*

If we hover over a cluster, we are also shown the start, stop, restart, lock, and delete icons which allow us to manage the cluster. In Figure A-6, by clicking the cluster, then we are taken to the cluster configuration screen which allows us to modify the properties of the cluster as well as access the SparkUI screen for a specific cluster.

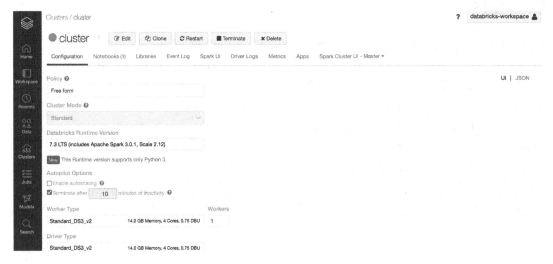

Figure A-6. *The cluster configuration screen*

In Figure A-7, we can see the advanced cluster configuration section which lets us modify any initialization scripts to the cluster, that is, scripts that are run when the cluster starts up, which lets us do things like ensure the .NET runtime is installed.

Figure A-7. *Cluster advanced configuration options*

In Figure A-8, we can see the notebooks which are attached to the cluster, and then in Figure A-9, we can see the Spark UI for the cluster. One useful thing to note is that even if the cluster has been shut down, you can still see the Spark UI; however, once a cluster has been terminated, the data is gone.

Figure A-8. *The notebooks connected to the cluster*

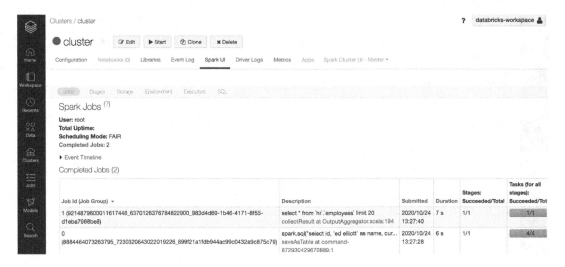

Figure A-9. *The Spark UI; note that the cluster has shut down, but the Spark UI is still available*

In Figure A-10, we can see any logging from the Apache Spark jobs that ran on the cluster. If you wrote anything to the console in your .NET for Apache Spark job, then the output would be on this screen. This is particularly useful if your application crashes and you output a stack trace; this is where you will see the error details.

Figure A-10. *Message logging and error messages are displayed on the "Driver Logs" tab of the cluster configuration screen*

Jobs

The next screen is the Jobs screen which is where we can run our .NET for Apache Spark applications, either on a schedule or on demand. When you click the Jobs screen, you can see a list of the jobs that have been created and click the "Create Job" button. We can see in Figure A-11, we have a single job configured.

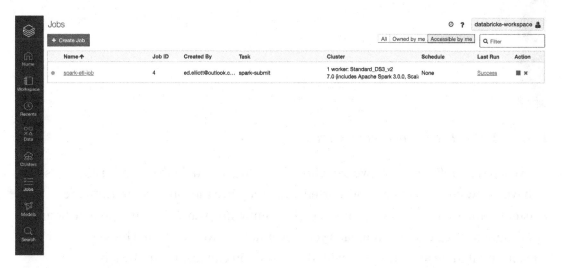

Figure A-11. *The "Jobs" screen*

When you click a Job, you enter the Job screen, as we can see in Figure A-12, which shows us the previous runs of the Job, including the output logs and the Spark UI for each execution of the job, no matter whether the cluster is still available or not. We can also see the configuration for the Job which includes the parameters which we must pass to spark-submit and the type of cluster we want to use.

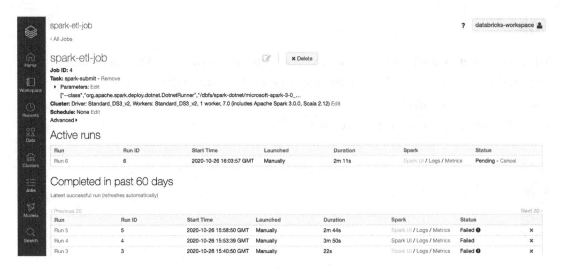

Figure A-12. *The Job overview screen*

When we configure a job, we can choose to either run the job on an existing cluster or have a new cluster created and started and then have the job executed and the cluster terminated at the end. This is a really compelling way to execute Apache Spark applications in the cloud as it means you never have to worry about a broken VM or cluster, and if there are any issues with the hosts, you can just rerun the job.

The way we configure our .NET for Apache Spark applications on Databricks is well documented by Microsoft: `https://docs.microsoft.com/dotnet/spark/tutorials/databricks-deployment`.

The steps to deploy your application in brief are to configure the cluster with an init script that deploys the .NET worker application so UDFs can work and then upload your published application, the .NET worker, and the init script to DBFS which is a shared file system accessible by Databricks. When you have everything uploaded, then you can create a Job that points to the .NET for Apache Spark JAR file and the zip containing your published application, and the .NET for Apache Spark application will run on the Databricks, Apache Spark cluster.

In Figure A-13, we can see a .NET for Apache Spark Job being created by manually specifying the spark-submit parameters.

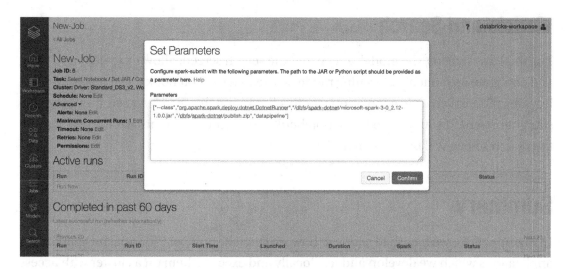

Figure A-13. *Manually specifying the spark-submit parameters*

Once we have created our job parameters, we can also create a schedule to run the job, as we can see in Figure A-14.

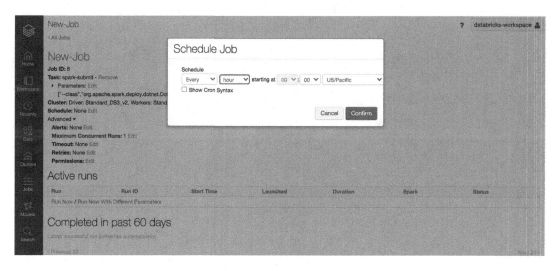

Figure A-14. *Specifying the parameters for the Job*

If you do not want to schedule your Job, you can trigger it manually in the UI or use the "databricks-cli" to run the job; you can even change the parameters that you are passing in. The Microsoft page that describes how to deploy and run a .NET for Apache Spark job on a Databricks cluster also shows how to configure the databricks-cli.

Models and Search

The final two screens, the model screen and the search screen, allow us to deploy and configure any machine learning models or search the notebooks. The machine learning models screen is out of scope for this book, and the search screen should be straightforward. If you want to find a notebook which contains some specific text, then the search screen will help you to find it.

Summary

In this appendix, we have had a look at how we can take our .NET for Apache Spark applications which we develop and test locally and execute them on a cluster with access to the data we need with the scale and elasticity to execute our jobs efficiently in the cloud using Databricks.

APPENDIX B

Implementing .NET for Apache Spark Code

The nature of the .NET for Apache Spark project is that the Apache Spark team releases some functionality, and then the .NET team has to replicate that functionality in .NET for Apache Spark. This appendix will show you how to implement calls to objects and methods in Apache Spark that are not yet available in .NET for Apache Spark. We will then show how to contribute your fixes back to the community and the .NET for Apache Spark project. To see the contributing guidelines for the project, see `https://github.com/dotnet/spark/blob/master/CONTRIBUTING.md` and see my tips for how to get your changes accepted efficiently at the end of the appendix.

There are roughly four categories of changes that you might want to make to the code in .NET for Apache Spark:

1. Add a method to an existing object.

2. Add a new object.

3. Add a new type to the serializer/deserializer.

4. Change the core framework.

We will be focusing on the first three types of change, and they increase in difficulty from adding a new method on an existing object to adding a new data type to the proxying code between the .NET and the Java VM code.

Making changes to the .NET for Apache Spark project will require both git and C# knowledge which are outside of the scope for this book.

© Ed Elliott 2021
E. Elliott, *Introducing .NET for Apache Spark*, https://doi.org/10.1007/978-1-4842-6992-3

Setting Up Your Build Environment

The .NET for Apache Spark project has some excellent documentation on setting up your build environment either on Windows or on Linux/macOS, and you can read the documentation at *https://github.com/dotnet/spark/tree/master/docs/building*. The key steps you need to follow are:

- Fork the spark/dotnet GitHub repo.

- Clone your fork to your developer machine.

- Install a JDK.

- Install Maven.

- Install the .NET SDK or install Visual Studio.

If we look at the current folder structure in the git repo, we have the folders described in Table B-1.

Table B-1. *The folder structure in the .NET for Apache Spark code repo*

Folder	Description
benchmark	This is the code to build and run the TPC-H benchmarks, which are a critical way to measure any data access system's performance
deployment	Documentation on various deployment options such as Databricks, AWS EMR, and Azure HD Insights
dev	Contains the scalafmt.conf file for the Scala development
docs	Documentation, a getting started guide as well as development documentation and coding guidelines
eng	Build properties and configuration files
examples	Various example projects showing how to use .NET for Apache Spark
script	Scripts used for releasing the project
src	This is the source code. It contains two distinct paths, csharp and scala

To build the csharp code, we can either run the build.sh/build.cmd file, depending on whether we are in Linux/macOS or Windows, in the root of the directory, which will build the code, or we can open the src/csharp/Microsoft.Spark.sln file using Visual

Studio or another IDE that can build C# projects such as JetBrains Rider. When you build the solution, either on the command line or via an IDE, all of the output files are built into the artifacts directory from the root directory. The artifacts directory will not exist when you first clone the repo.

To build the scala code, we change to the spark/src/scala directory and use the maven command "mvn clean package" to build the JAR files needed for the project. You can also use a scala IDE such as JetBrains' IntelliJ to open the project and to run the maven commands to clean and package the code into JAR files.

At this stage, we should be able to build the code, and so we can move on to what is involved with implementing our own objects and methods.

Adding a Method to an Existing Object

The simplest case is to add a method to an existing object. In this example, we will look at how to add a method to an existing object. The example we will use is this pull request, which was to add the "Table" method to the "DataFrameReader" (*https:// github.com/dotnet/spark/pull/665/files*). The DataFrameReader object in scala implemented the table method, and the documentation is at *https://spark.apache. org/docs/latest/api/java/org/apache/spark/sql/DataFrameReader.html#table- java.lang.String.* However, the DataFrameReader in .NET for Apache Spark did not implement the method. In this case, we can see there is a method we would like to use in Apache Spark, but it is not available in .NET for Apache Spark.

If there is a method we would like to implement, we first need to know where to implement it, and so we search the .NET for Apache Spark codebase for the object where the method belongs. In this case, the DataFrameReader sits in the src/csharp/Microsoft. Spark/Sql folder.

The first step is to translate the method's name from the scala naming standard, which is camel case, that is, lowercase first letter and then uppercase first letter for each word into the C# standard, pascal case, that is, uppercase first letter and of each first letter of each word. Here, we change "table" to "Table". We then need to convert any types from the scala types to .NET types. The definition for the Apache Spark method is `public Dataset<Row> table(String tableName)`, so we need to switch `String` to `string` and `Dataset<Row>` to `DataFrame`.

Once we have our method signature, we use the already defined _jvmObject, which is a reference to the object in the Java VM to invoke the method we would like to invoke in the Java VM. In this case, the table method is passing in any parameters we need.

Because the table method returns a new DataFrame, we need to, on the .NET side, also create a new DataFrame and store a reference to the Java VM instance of the new DataFrame. We can then return the new DataFrame to the .NET caller, and we are done. Listing B-1 shows the .NET code for the new method.

Listing B-1. The .NET implementation of the new method

```
public DataFrame Table(string tableName) =>
    new DataFrame((JvmObjectReference)_jvmObject.Invoke("table",
    tableName));
```

If we built the project, we could now reference the new method from a DataFrameReader.

Adding a method to an existing object such as the DataFrameReader is the simplest way to add any missing functionality you might need.

Adding a New Object

The next type of change is where you want to add a whole new object. In this example, we will look at the Table object that is returned by the spark catalog in calls such as listTables and getTable. The code for the .NET version of Table can be found at https://github.com/dotnet/spark/blob/master/src/csharp/Microsoft.Spark/Sql/Catalog/Table.cs.

When we create a new class, we need to implement the interface IJvmObjectReferenceProvider, which allows our new class to be created from a Java VM reference. We also create an internal constructor that is going to be passed in a JvmObjectReference. This is the link between the .NET instance of the class and the Java VM instance of the class. If you create ten objects, then there will be ten specific JvmObjectReferences.

If the scala version of the object also includes public constructors, then we would add them as well. In this example, Tables are only constructed by Apache Spark, and there are no public constructors, but consider something like the CountVectorizer (*https://github.com/dotnet/spark/blob/master/src/csharp/Microsoft.Spark/ML /Feature/CountVectorizer.cs*) which has two public constructors, so we replicate the public constructors as well as the internal-only constructor with the JvmObjectReference.

When we implement IJvmObjectReferenceProvider, we need to provide a way to return the _jvmObject, as we do this using this method group: JvmObjectReference IJvmObjectReferenceProvider.Reference => _jvmObject;.

At this point, we should have an object that implements IJvmObjectReferenceProvider and has an internal constructor that is passed the JvmObjectReference, and we store that object reference, likely as _jvmObject. The next task is to add the method, as we did in the previous section. We convert the naming standard to .NET and the Java VM types to .NET types and can call the method via the _jvmObject. Listing B-2 shows the call to the "database" method on the "Table" object.

Listing B-2. Calling the database method on the Table object

```
public string Database => (string)_jvmObject.Invoke("database");
```

The interesting thing to note here is that the return type from the method is a native type, a string, so there is no need to create a new Java VM reference to the string. We just have the string, so we need to cast from object to string.

Hopefully, at this point, you can see how to add a new method to an existing object as well as see some of the plumbing that is required for new objects. The next section deals with the case that the current serializer/deserializer does not support the data type you need to either pass across to the Java VM or to receive back as the return type, and we need to add a new type.

Adding a New Data Type to the Serializer/ Deserializer

The less common type of change we need to make is where we either need to pass a parameter to Apache Spark or receive the result back as a type of parameter that is currently not supported by the serializer/deserializer. If you try to implement a method that has an unsupported parameter type, you will encounter this exception "Type {0} not supported yet".

In order to implement the type, we need to understand how the process of sending requests to Apache Spark and getting the responses works. If we recall from the way .NET for Apache Spark works, requests are proxied from .NET to the Java VM using a local network connection. Then the serialization process needs to understand all of the data types that are being passed across in either direction.

To send a request from .NET to the Java VM, the .NET code builds a payload, which is a network packet containing the request that can be read by the Java VM.

The process from a code point of view is to send a request from .NET to Java; we follow this flow:

- The class, such as `DataFrame`, calls `JvmObjectReference.Invoke`.

- Invoke calls `JvmBridge.CallNonStaticJavaMethod` or `CallStaticJavaMethod`.

- The network payload is created using `PayloadHelper.BuildPayload`.

- The request, including the method name, is added to the payload.

- Each argument is written to the payload; each data type has an identifier. For example, an int32 is "i", and an int64 is "g". I don't believe there is a system for deciding which letter to use. It just needs to be a letter that hasn't been used before.

- After the type identifier is written to the payload, the actual type is converted to a series of bytes, depending on the type, and then written to the payload.

- The payload is sent to the Java VM.

The .NET for Apache Spark scala code then received the payload and followed this flow:

- The `DotnetBackendHandler.handleBackendRequest` function is called to read the payload from the network connection.

- If it is a method call, instead of something like destroying an object or removing a thread, then `handleMethodCall` is called.

- `handleMethodCall` reads the network stream until all of the parameters have been converted from bytes into their appropriate type.

- To convert from bytes into the type, the type identifier is read, and then the appropriate function is called, such as readInt if the type identifier is "i".

Once the payload has been reversed into the name of the function to call and the parameters in the native types, the .NET for Apache Spark scala code calls the Apache Spark code and waits for the response. If the response includes a return value, the scala code creates its payload using the same system that the .NET code uses. That is, the scala code:

- Writes the payload header and then the type of the response to the payload.

- The response is written as a series of bytes.

- The response is sent back to .NET.

The .NET code then does the same thing that the scala code did when it received the request. The .NET code reads the payload and, depending on the type of the response, reads the response as a series of bytes into the appropriate type.

The type of parameters being sent can be a native type such as an int or a string, or it can be a more complex object such as an ArrayType, MapType, or a JvmReference, which will need to be converted to an object on the .NET side or used to call specific methods on the Java VM side.

Let's walk through an example of how a new type was added to .NET for Apache Spark. In this example, I was adding support for the Bucketizer; however, this relied on a type that is defined, such as double[][] that is an array of double arrays, and that type was not supported by the serializer/deserializer process.

The first thing to do is to add the type identifier to the payload helper which is in *src/csharp/Microsoft.Spark/Interop/Ipc/PayloadHelper.cs* where the type identifiers are coded as static bytes as shown in Listing B-3.

Listing B-3. The data type identifiers for proxying between .NET and the Java VM

```
private static readonly byte[] s_int32TypeId = new[] { (byte)'i' };
private static readonly byte[] s_int64TypeId = new[] { (byte)'g' };
private static readonly byte[] s_stringTypeId = new[] { (byte)'c' };
private static readonly byte[] s_boolTypeId = new[] { (byte)'b' };
private static readonly byte[] s_doubleTypeId = new[] { (byte)'d' };
private static readonly byte[] s_dateTypeId = new[] { (byte)'D' };
private static readonly byte[] s_timestampTypeId = new[] { (byte)'t' };
```

```
private static readonly byte[] s_jvmObjectTypeId = new[] { (byte)'j' };
private static readonly byte[] s_byteArrayTypeId = new[] { (byte)'r' };
private static readonly byte[] s_doubleArrayArrayTypeId = new[] { (byte)'A' };
```

The type identifier for the double[][] is last in Listing B-3 and is an uppercase "A".

In the PayloadHelper class, there is a method that converts each argument type to a stream of bytes. The method ConvertArgsToBytes includes a large switch statement, and we need to add a case for our new type. In Listing B-4, we show the case statement that allows a double[][] to be written to the payload.

Listing B-4. The .NET case statement to write the double[][] to the payload

```
case double[][] argDoubleArrayArray:
    SerDe.Write(destination, s_doubleArrayArrayTypeId);
    SerDe.Write(destination, argDoubleArrayArray.Length);
    foreach (double[] doubleArray in argDoubleArrayArray)
    {
        SerDe.Write(destination, doubleArray.Length);
        foreach (double d in doubleArray)
        {
            SerDe.Write(destination, d);
        }
    }
    break;
```

The changes required to the PayloadHelper to add the double[][] can be viewed as a github diff (*https://github.com/dotnet/spark/pull/378/files?file-filters%5B%5D=.cs#diff-fde5da3dcd720ed2407b0b222a9cd2ae4f7d7305f34a8023b0c8 8282ccc7cffe*).

Logically, the next thing to do is to switch to the scala code and deserialize the type that we are writing to the payload. The first thing to note with the scala code is the way the project is laid out; there is a version of the scala code for each version of Apache Spark. So, at this point in time, there is a src/Microsoft-spark-2-3, src/Microsoft-spark-2-4, src/Microsoft-spark-3-0, and all three contain a copy of the same code. The easiest way to work with this is to get one version working as you want and then copy the changes to the other two versions.

In the scala code, in the SerDe object, we can see the readList function which converts the payload data into an array type by reading the type identifier and then calling the appropriate function for that type. In Listing B-5, we can see the lookup from the type identifier to the function that can read that type.

Listing B-5. The type identifier defines which function to call, in our case, readDoubleArrArr

```
def readList(dis: DataInputStream): Array[_] = {
    val arrType = readObjectType(dis)
    arrType match {
      case 'i' => readIntArr(dis)
      case 'g' => readLongArr(dis)
      case 'c' => readStringArr(dis)
      case 'd' => readDoubleArr(dis)
      case 'A' => readDoubleArrArr(dis)
      case 'b' => readBooleanArr(dis)
      case 'j' => readStringArr(dis).map(x => JVMObjectTracker.
      getObject(x))
      case 'r' => readBytesArr(dis)
      case _ => throw new IllegalArgumentException(s"Invalid array type
      $arrType")
    }
}
```

We then need to define the method that understands how to read a double[][], which is shown in Listing B-6.

Listing B-6. The readDoubleArrArr function

```
def readDoubleArrArr(in: DataInputStream): Array[Array[Double]] = {
    val len = readInt(in)
    (0 until len).map(_ => readDoubleArr(in)).toArray
}
```

At this point we have, we can create a double[][] in our own code and pass it as a parameter to the .NET code, which can write the double[][] over to the Java VM by serializing the parameter, and then the scala code that runs on the Java VM can read that parameter and pass it to an Apache Spark method to work with. What we need to do now is the reverse; we need to write the scala code to serialize any responses that are of our new type and then write the .NET code that reads from the payload and converts it into a .NET type that can be passed back to the calling application.

Back in the SerDe object in scala, we need to add our type identifier to writeType as shown in Listing B-7.

Listing B-7. Writing the type identifier back to .NET from scala

```
def writeType(dos: DataOutputStream, typeStr: String): Unit = {
typeStr match {
  case "void" => dos.writeByte('n')
  case "character" => dos.writeByte('c')
  case "double" => dos.writeByte('d')
  case "doublearray" => dos.writeByte('A')
  case "long" => dos.writeByte('g')
  case "integer" => dos.writeByte('i')
  case "logical" => dos.writeByte('b')
  case "date" => dos.writeByte('D')
  case "time" => dos.writeByte('t')
  case "raw" => dos.writeByte('r')
  case "list" => dos.writeByte('l')
  case "jobj" => dos.writeByte('j')
  case _ => throw new IllegalArgumentException(s"Invalid type $typeStr")
  }
}
```

Then in the writeObject function, we add a case statement for a double[][] which relies on internal JVM knowledge to know that the class name is "[[D". If you look at the other items in the match statement, you can figure out the type. In Listing B-8, we see the case matching the double[][] type.

Listing B-8. Writing the double[][] back out to .NET from scala

```
case "[[D" =>
  writeType(dos, "list")
  writeDoubleArrArr(dos, value.asInstanceOf[Array[Array[Double]]])
```

Finally, in the scala code, we need to write our function to physically write the contents of the double[ ][ ] out to the payload which we see in Listing B-9.

Listing B-9. Writing the double[][] back out to the payload to be written back to .NET

```
def writeDoubleArrArr(out: DataOutputStream, value: Array[Array[Double]]):
Unit = {
 writeType(out, "doublearray")
 out.writeInt(value.length)
 value.foreach(v => writeDoubleArr(out, v))
}
```

To see all the changes which you need to make to the SerDe.scala file, see https:// github.com/dotnet/spark/pull/378/files#diff-17d440262c6de3c131abc642f9bf32a c608b9c5e0b592b9f69d0aec1d6db2740.

The last thing to do then is to go back to the .NET csharp code and write the code that can read the double[ ][ ] from the payload and convert it back into a .NET type. In *src/csharp/Microsoft.Spark/Interop/Ipc/JvmBridge.cs*, there is a switch statement that has a case for each of the type identifiers. In Listing B-10, we can see the additional case for the new data type.

Listing B-10. Reading the double[][] back from the payload and writing it to a new double[][] that we create

```
case 'A':
    var doubleArrayArray = new double[numOfItemsInList][];
    for (int itemIndex = 0; itemIndex < numOfItemsInList; ++itemIndex)
    {
        doubleArrayArray[itemIndex] = ReadCollection(s) as double[];
    }
    returnValue = doubleArrayArray;
    break;
```

The full change for the JvmBridge class can be seen at *https://github.com/dotnet/ spark/pull/378/files?file-filters%5B%5D=.cs#diff-7cba32304e5f3d61b3732439a5 7fbe3f46908468ca6d56e69983cd49880afc41.*

Now that both the .NET code and the scala code can read and write the new data type, you are free to pass it in as a parameter and use it as a return value which we can see with the Bucketizer as in Listing B-11.

Listing B-11. Using the new data type in the .NET for Apache Spark project

```
public Bucketizer SetSplits(double[] value) => WrapAsBucketizer(_jvmObject.
Invoke("setSplits", value));

public double[][] GetSplitsArray() => (double[][])_jvmObject.
Invoke("getSplitsArray");
```

Tips to Help Get Contributions Merged

At this point, you should be able to add a missing method, add a missing object, and even add extra data types to the serialization/deserialization process, which should allow you to extend the .NET for Apache Spark project should there be some missing functionality that is a blocker for you, but because the project is open source, you also have the option to create a pull request and submit the new code to the team who can accept the pull request and merge the code into the project, which benefits you because you do not have to maintain your own version and benefits other users as they can use the functionality as well.

This section will cover some common issues that need to be corrected before the pull request can be accepted.

Formatting

The most common issue is formatting. The requirements for any code merged in are very specific. In this section, we will look at the most common formatting issues. The formatting standards are the same as the dotnet/runtime project. The .NET for Apache Spark C# coding style is documented at

https://github.com/dotnet/spark/blob/master/docs/coding-guidelines/csharp-coding-style.md

And the dotnet/runtime standards are documented here:

https://github.com/dotnet/runtime/blob/master/docs/coding-guidelines/
coding-style.md

Line Length

The maximum amount of characters used to be 100 but has recently been increased to 110, and if your code is longer than 110 characters, then you will need to format it so that it fits within the maximum width.

Listing B-12 gives an example of a too long line and then a separate version of the line that has been formatted so that it is within the 100-character limit.

Listing B-12.   The first line is too long and will fail a code review; the second is within the character limit

```
//Too Long
        public IEnumerable<string> GetInputCols() => ((string[])
        (_jvmObject.Invoke("getInputCols"))).ToList();

//Formatted so it is within 100 characters
        public IEnumerable<string> GetInputCols() =>
            ((string[])(_jvmObject.Invoke("getInputCols"))).ToList();
```

Comments

Every public method requires a comment header, which includes a summary of how the method works, which can be copied from the Apache Spark documentation as a start and then expanded upon as necessary. The comment header also contains a description of each of the parameters and the details of the return value, which typically shows the type and the detail of what it is returned. The comment header is in a specific format as it is used to generate the documentation on the Microsoft Docs site (*https://docs. microsoft.com/en-us/dotnet/api/?view=spark-dotnet*).

Listing B-13 shows an example of a correct comment header.

Listing B-13. A valid comment header

```
/// <summary>
/// Negate the given column.
/// </summary>
/// <param name="self">Column to negate</param>
/// <returns>New column after</returns>
```

The important things to note are that each parameter has an appropriate `<param name=` tag and the name is correct. The contents of the `<summary>` tag also must end with a full stop, in that the summary is always a valid sentence. If the description of a parameter or the return value would put the character limit over 100, then you can use a multiline version of the comment header as shown in Listing B-14.

Listing B-14. Comment tags that would go over 110 characters can be multiline tags

```
/// <summary>
/// Negate the given column.
/// </summary>
/// <param name="self">
/// Parameter description which would be over 110 characters wide
/// </param>
/// <returns>
/// Return description which would be over 110 characters wide
/// </returns>
```

Copyright Notice

Every new code file must have this copyright notice at the top. This includes test files. The copyright notice is shown in Listing B-15.

Listing B-15. The copyright notice at the top of every code file

```
// Licensed to the .NET Foundation under one or more agreements.
// The .NET Foundation licenses this file to you under the MIT license.
// See the LICENSE file in the project root for more information.
```

Extra Blank Lines

There are no cases where two blank lines next to each other are acceptable, and there must be a blank line at the end of the code file. Listing B-16 shows an example of an unacceptable extra line.

Listing B-16. Extra empty lines are not acceptable

```
Assert.Equal(expectedUid, bucketizer.Uid());

DataFrame input = _spark.Sql("SELECT ID as input_col from range(100)");
```

Explicit Type Declarations

One place where the .NET for Apache Spark project diverges from the dotnet standard is over the use of "var" to declare a variable rather than declaring the data type.

"var" must only be used when the return type is explicitly declared; otherwise, you must declare the type. Listing B-17 shows some examples of when var is and is not acceptable.

Listing B-17. Use of var over explicit types

```
var foo = new Foo(); // OK
Foo foo = new Foo(); // NOT OK

var bar = foo as Bar; // OK
Bar bar = foo as Bar; // NOT OK

var bar = (Bar)foo; // OK
Bar bar = (Bar)foo; // NOT OK

string str = "hello"; // OK
var str = "hello"; // NOT OK
int i = 0; // OK
var i = 0; // NOT OK
```

```
var arr = new string[] { "abc", "def" }; // OK
string[] arr = new[] { "abc", "def" }; // NOT OK
var arr = new[] { "abc", "def" }; // NOT OK

string str = foo.GetString(); // Function name shouldn't matter.
var str = foo.GetString(); // NOT OK
```

In some circumstances, you may think it is possible to use var, but be careful. Listing B-18 shows where a Bucketizer is created using new, which would allow var to be used. Then SetInputCol called, which also returns a Bucketizer, but because it isn't the original object that is returned, we need to switch to the explicit type.

Listing B-18. Explicit data types, even when they are pretty obvious

```
var bucketizer = new Bucketizer(); //OK
var bucketizer = new Bucketizer().SetInputCol("col"); //NOT OK
```

Tests

The last thing that can help a pull request get merged faster is the unit and integration tests being added correctly. The same coding standards apply to the tests as well as the code, so be careful, especially with the use of var.

The general idea of the tests that validate that Apache Spark methods are called correctly is to ensure that the method exists, and it takes the correct parameters and returns the right type. The idea is not to validate that Apache Spark works correctly as that is down to the Apache Spark project, and if the behavior of a method changes, then the .NET for Apache Spark project's tests could start failing.

View the file at the following location to find examples on how to test methods and verify that the correct data type is returned:

src/csharp/Microsoft.Spark.E2ETest/IpcTests/Sql/FunctionsTests.cs

Listing B-19 shows some examples from this file.

Listing B-19. The tests for the Functions.Sum and Functions.SumDistinct

```
Assert.IsType<Column>(Sum(col));
Assert.IsType<Column>(Sum("col"));

Assert.IsType<Column>(SumDistinct(col));
Assert.IsType<Column>(SumDistinct("col"));
```

Summary

In this appendix, we have covered what is involved in extending the .NET for Apache Spark project, in case you are blocked requiring specific functionality that is available in Apache Spark but has yet to be ported to .NET for Apache Spark. We also looked at what is involved in contributing any code you write back to the project.

Index

259

© Ed Elliott 2021
E. Elliott, *Introducing .NET for Apache Spark*, https://doi.org/10.1007/978-1-4842-6992-3